On Dynamics of Hamiltonian Systems on Infinite Lattices

Dissertation

zur Erlangung des akademischen Grades
doctor rerum naturalium
(Dr. rer. nat.)
im Fach Mathematik

eingereicht an der
Mathematisch-Naturwissenschaftlichen Fakultät II
der Humboldt-Universität zu Berlin

von
Dipl.-Math. Dipl.-Ing. Carsten Patz
geboren am 17.10.1976 in Wertheim am Main

Präsident der Humboldt-Universität zu Berlin
Prof. Dr. Jan-Hendrik Olbertz

Dekan der Mathematisch-Naturwissenschaftlichen Fakultät II
Prof. Dr. Elmar Kulke

Gutachter:
1. Prof. Dr. Alexander Mielke
2. Prof. Dr. Etienne Emmrich
3. Prof. Dr. Michael Herrmann

eingereicht am: 30. September 2013
Tag der mündlichen Prüfung: 24. Februar 2014

Bibliographic information published by the Deutsche Nationalbibliothek

The Deutsche Nationalbibliothek lists this publication in the Deutsche Nationalbibliografie; detailed bibliographic data are available in the Internet at http://dnb.d-nb.de .

ISBN 978-3-8325-3916-0

Logos Verlag Berlin GmbH
Comeniushof, Gubener Str. 47,
10243 Berlin
Tel.: +49 (0)30 42 85 10 90
Fax: +49 (0)30 42 85 10 92
INTERNET: http://www.logos-verlag.de

Vorwort

Die vorliegende Arbeit entstand unter der Betreuung von Prof. Dr. Alexander Mielke in Stuttgart und Berlin sowie während eines Forschungsaufenthaltes bei Prof. Dr. Claude Le Bris an der École des Ponts bei Paris. Mein Dank gilt zunächst ganz besonders Alex für die Einführung in diese spannende Themenfeld, die fortlaufende Unterstützung und die große Geduld bei unseren Diskussionen, Claude ebenfalls für die zahlreichen Gespräche und die Unterstützung meines Aufenthalts in Paris und natürlich Olga Kuphal vom WIAS für die unkomplizierte Lösung aller anfallenden organisatorischen Themen.

Im Verlauf der Arbeit hatte ich mit vielen Personen intensiv diskutiert und habe dabei unglaublich viel gelernt. In Berlin sind mir am WIAS Dr. Annegret Glitzky, mit der ich lange das Büro geteilt habe, Prof. Dr. Anton Bovier, Dr. Christiane Kraus sowie Dr. Marita Thomas und Dr. Florian Schmid mit denen ich bereits in Stuttgart studiert hatte, und nicht zuletzt Prof. Dr. Michael Herrmann von der HU Berlin, besonders in Erinnerung geblieben. In meiner Zeit in Paris habe ich mit Dr. Frederic Legoll und Dr. Xavier Blanc und weiteren Wissenschaftlern am CERMICS zusammen gearbeitet. Ich danke allen für diese Zeit!

Der Grundstein für diese Arbeit wurde während meines Studiums in Stuttgart gelegt. Hier bedanke ich mich bei Dr. Vanessa Miemietz und Markus Bez für die intensiven Diskussionen sowie Prof. TeknD Timo Weidl und Prof. Dr. Jörg Brüdern für die spannenden Vorlesungen und nicht zuletzt Georg Feldmeyer, meinem Mathematiklehrer in der Oberstufe, der meine Begeisterung für das Fach nachhaltig geweckt hat.

Der größte Dank gilt schließlich meinen Eltern und meiner Frau Lisa, die mich so ausdauernd unterstützt und mir alles ermöglicht haben sowie meinen Kindern Paula und Moritz, die an vielen Wochenenden und Abenden auf mich verzichten mussten.

Carsten Patz
Stuttgart, Dezember 2014

Summary

The thesis at hand consists of four individual papers on questions concerning the interaction of infinitely many particles. The prototype of the potential energy that determines the interaction in the one-dimensional case reads as follows $E_{\text{pot}}\big((x_j)_{j\in\mathcal{J}}\big) = \sum_{j\in\mathcal{J}}\big(\sum_{1\leq k\leq K} V_k(x_j - x_{j-k}) + W(x_j)\big)$, where $x_j \in \mathbb{R}$ denote the positions and $\mathcal{J}$ a large finite or infinite discrete set. This class of systems includes the celebrated Fermi-Pasta-Ulam (FPU) chain and the discrete Klein-Gordon (DKG) equation.

In the first paper, *Dispersive stability of infinite-dimensional Hamiltonian systems on lattices*, the main result is that if the nonlinearity is of degree $\beta > 4$, then small localized initial data decay just as in the linear case. The proof relies on similar arguments as analogous results in the theory of dispersive PDE and sharp ℓ^p-decay estimates for the linearized problem using oscillatory integrals. The latter is crucial since this avoids the nonoptimal interpolation between different ℓ^p spaces.

The second paper, *Uniform asymptotic expansions for the infinite harmonic chain*, refines the linear theory in the one-dimensional case. An asymptotic expansion for the Green's function matrix of the linearized system is derived that holds uniformly in space and also resolves the microscopic oscillations. As a consequence we find that the ℓ^p-bounds are indeed sharp. The asymptotic behaviour as $t \to \infty$ is determined by three regions: wave fronts that decay like $\sim t^{-1/3}$, a region between the fronts $\sim t^{-1/2}$ and a region of exponential decay outside the outer wave fronts. The fronts correspond to wavenumbers where the dispersion relation degenerates and match to an Airy function by choosing the correct scaling.

For nonlinearities up to order $\beta < 3$ it is known that there exist solitary waves even for small and localized initial conditions. The third paper *Propagation of small amplitude wave fronts in infinite oscillator chains* aims to close the gap for $\beta \in [3, 4]$ using numerical experiments and formal calculations. It turns out that in case $\beta = 3$ the wave fronts still decay like $\sim t^{-1/3}$, but this time the profile matches to the solution of a nonlinear ODE, namely a Painlevé equation. For $\beta \in (3, 4]$ there is a crossover between linear and nonlinear behaviour at the front. Simulation results suggest that there are no solitary waves as long the initial conditions are sufficiently small.

Finally, in the last paper, *Finite-temperature coarse-graining of one-dimensional models*, the authors present a possible approach for the computation of free energies and ensemble averages of one-dimensional coarse-grained models in materials science. The approach is based on a thermodynamic limit process, and makes use of ergodic theorems and large deviations theory. In addition to providing an efficient computational strategy for ensemble averages, the approach allows for assessing the accuracy of approximations commonly used in practice.

Zusammenfassung

Die vorliegende Arbeit besteht aus vier einzelnen Artikeln, die Fragen über die die Wechselwirkung von Systemen von unendlich vielen Teilchen adressieren. Im eindimensionalen Fall hat die potentielle Energie, die Wechselwirkung bestimmt, die Form $E_{\mathrm{pot}}\big((x_j)_{j\in\mathcal{J}}\big) = \sum_{j\in\mathcal{J}}\big(\sum_{1\leq k\leq K} V_k(x_j - x_{j-k}) + W(x_j)\big)$. Hierbei bezeichnet $x_j \in \mathbb{R}$ die Positionen und $\mathcal{J}$ eine große oder unendliche diskrete Menge. Diese Klasse von Systemen beinhaltet die bekannte Fermi-Pasta-Ulam (FPU) Kette sowie die diskrete Klein-Gordon (DKG) Gleichung.

Im ersten Artikel, *Dispersive stability of infinite-dimensional Hamiltonian systems on lattices*, ist das zentrale Ergebnis, dass kleine, lokalisierte Anfangsbedingungen analog zum linearen Fall mit der Zeit abklingen, sofern der Grad der Nichtlinearität $\beta > 4$ ist. Der Beweis verwendet ähnliche Argumente wie das analoge Resultat in der Theorie dispersiver PDGL und scharfe ℓ^p-Abschätzungen über oszillierende Integrale. Letzteres ist zentral, um nicht-optimale Interpolation zwischen ℓ^p-Räumen zu vermeiden.

Der zweite Artikel, *Uniform asymptotic expansions for the infinite harmonic chain*, vertieft die lineare Theorie im eindimensionalen Fall. Es werden bezüglich des Ortes gleichmäßige asymptotische Entwicklungen für die Greensche Funktion des linearen Systems hergeleitet, die auch mikroskopische Oszillationen auflösen. Hieraus folgt, dass die genannten ℓ^p-Abschätzungen in der Tat scharf sind. Das asymptotische Verhalten insgesamt teilt sich in drei Bereiche auf: Wellenfronten, die entsprechend $\sim t^{-1/3}$ abklingen, Regionen dazwischen mit $\sim t^{-1/2}$ sowie Bereiche vor den äußeren Wellenfronten mit exponentiellem Abklingen. Die Fronten entsprechen Wellenzahlen, an denen die Dispersionsrelation degeneriert. In einer korrekten Skalierung entspricht die Form der Fronten einer Airy-Funktion.

Bei Nichtlinearitäten bis $\beta < 3$ ist existieren auch für beliebig kleine, lokalisierte Anfangsbedingungen solitäre Wellen. Der dritte Artikel, *Propagation of small amplitude wave fronts in infinite oscillator chains*, versucht die sich daraus ergebende Lücke für $\beta \in [3, 4]$ zu schließen. Es zeigt sich, dass Wellenfronten im Falle $\beta = 3$ immer noch $\sim t^{-1/3}$ abklingen. Das sich herausbildende Profil entspricht nun allerdings der Lösung einer nichtlinearen DGL, einer Painlevé Gleichung. Für $\beta \in (3, 4]$ findet ein annähernd kontinuierlicher Übergang zwischen linearem und nichtlinearem Verhalten der Front statt. Simulationsergebnisse legen nahe, dass es hier bei hinreichend kleinen Anfangesbedingungen nicht zur Ausbildung von solitären Wellen kommt.

Im letzten Artikel, *Finite-temperature coarse-graining of one-dimensional models*, wird ein Verfahren zur Berechnung von freien Energien und Ensemble-Mitteln vorgeschlagen. Der Ansatz basiert auf dem thermodynamischen Limes und nutzt Ergodensätze und die Theorie über große Abweichungen. Zusätzlich wird eine effiziente Strategie zur Berechnung von Ensemble-Mitteln bereitgestellt. Dieser ermöglicht die Bewertung der Genauigkeit von in der Praxis häufig genutzten Ansätzen.

Contents

Chapter 1

Introduction

Large systems of particles interacting via the laws of classical mechanics are widely-used models in material science and statistical mechanics. In characterizing the behavior of such systems, the potential energy functional is essential. Restricting ourselves for the introduction, as well as in most of the upcoming chapters, to the one-dimensional case, a prototype looks like

$$E_{\text{pot}}\big((x_j)_{j\in\mathcal{J}}\big) = \sum_{j\in\mathcal{J}} \left(\sum_{1\le k\le K} V_k(x_j - x_{j-k}) + W(x_j) \right), \tag{1.1}$$

where the $x_j \in \mathbb{R}$ denote the positions of the atoms and either $\mathcal{J} = \{1, \ldots, J\}$ with $J \in \mathbb{N}$ or $\mathcal{J} = \mathbb{Z}$. In case of a finite number of particles suitable conditions have to be added for $j - k \le 0$ which we neglect for the moment. Conditions on the interaction potentials V_k and the on-site potential modeling the possible coupling to a background are also to be discussed later.

There is an enormous amount of literature on these kinds of systems. However, a major topic in this context is the relation between the dynamics of these large, but microscopic systems and a behavior that is in some sense macroscopic. This question was first posed in statistical physics more than 100 years ago and still is one of the most challenging fields of multi-scale analysis. In the context of statistical physics and thermodynamics, main questions are about the consistency of these two theories by taking the thermodynamic limit, i.e. increasing the number of particles in a suitable manner, see [Spo91, Rue69, Fis64]. Apart from theoretical considerations, computational aspects, that is computing macroscopic quantities based on microscopic models, attract interest.

A second category, that is far from thermodynamic fluctuations, is the reversive evolution of initial conditions that are well-defined on the microscopic level. Prototypical problems again concern the passage from discrete lattice dynamics to continuum models describing the effective dynamics on much larger spacial and/or temporal scales. Here, a special case is the long-time behavior of microscopic initial data. In this context, emergence and dynamics of coherent structures, e.g. solitary waves, are of particular interest, see for instance [Sco03, Ioo00]. The thesis at hand covers aspects of both categories. The latter is addressed in Chapter 2, 3 and 4. In Chapter 5 we study the thermodynamic limit for a system with potential energy of the form (1.1) and present a possible approach to compute free energies. In the next Sections 1.1 and 1.2 we give a short introductions to these topics.

1.1 Hamiltonian dynamics

Restricting the interaction to pure nearest-neighbor (NN) interaction and neglecting the on-site potential W, we obtain the prototype of Hamiltonian systems on lattices, namely the celebrated Fermi-Pasta-Ulam (FPU) chain. The Hamiltonian function results from (1.1) by adding the potential energy, i.e.

$$\mathcal{H}(\mathbf{x},\mathbf{p}) = \sum_{j\in\mathcal{J}} \left(\tfrac{1}{2}p_j^2 + V(x_j - x_{j-1})\right) ,$$

where $\mathbf{x} := (x_j)_{j\in\mathcal{J}}$ and $\mathbf{p} := (\dot{x}_j)_{j\in\mathcal{J}}$. Assuming $J = \mathbb{Z}$ the equations of motion in Newtonian form read as

$$\ddot{x}_j(t) = V'(x_{j+1} - x_j) - V'(x_j - x_{j-1}) , \qquad j\in\mathbb{Z} . \tag{1.2}$$

Further famous examples are the discrete Klein-Gordon chain and the discrete nonlinear Schödinger equation,

$$\begin{aligned} \ddot{x}_j(t) &= x_{j+1} - 2x_j + x_{j-1} + W'(x_j) , & j&\in\mathbb{Z} \\ \mathrm{i}\dot{u}_j(t) &= u_{j+1} - 2u_j + u_{j-1} + \alpha|u_j|^{\beta-1}u_j , & j&\in\mathbb{Z} . \end{aligned}$$

Here the potentials V and W are assumed to have the forms $V'(r) = ar + \mathcal{O}(r^\beta)$ and $W'(x) = bx + \mathcal{O}(x^\beta)$ as $r\to\infty$. I.e. $\beta > 1$ is used to measure the nonlinearity. Occasionally we will write the system under consideration in abstract form,

$$\dot{\mathbf{z}}(t) = \mathcal{L}\mathbf{z} + \mathcal{N}(\mathbf{z}) \tag{1.3}$$

with $\mathcal{L}$ being a linear operator on Banach spaces $\ell^p = \ell^p(\mathbb{Z};\mathbb{R}^2)$ and $\mathcal{N}$ concentrating all nonlinear terms. In many aspects the discrete Klein-Gordon and nonlinear Schrödinger equation show behavior similar to the FPU in (1.2). One main difference relates to the fact that the FPU is Galilean invariant, i.e. the transformation $(\mathbf{x},\mathbf{p}) \to (x_j + \xi + ct, p_j + c)_{j\in\mathbb{Z}}$ leaves (1.2) invariant. This point will be discussed in the upcoming chapters. For now we focus on the FPU chain.

The linearized FPU chain exhibits plane waves solutions of the form $x_j(t) = \mathrm{e}^{\mathrm{i}(\theta j + \hat{\omega} t)}$ if and only if $\hat{\omega} = \omega(\theta)$, where $\omega(\theta)^2$ is a trigonometric polynomial. Thus in general group velocity $c_{\mathrm{gr}}(\theta) = \omega'(\theta)$ and phase velocity $c_{\mathrm{ph}}(\theta) = \omega(\theta)/\theta$ do not match which leads to dispersive effects. As a consequence, localized solutions of the linear system decay in time as illustrated in Figure 1.1. In fact, at the wave fronts the solutions decay like $\sim t^{-1/3}$ while in the inner region we have $\sim t^{-1/2}$. The proof relies on applying methods from asymptotic analysis to the explicit solution

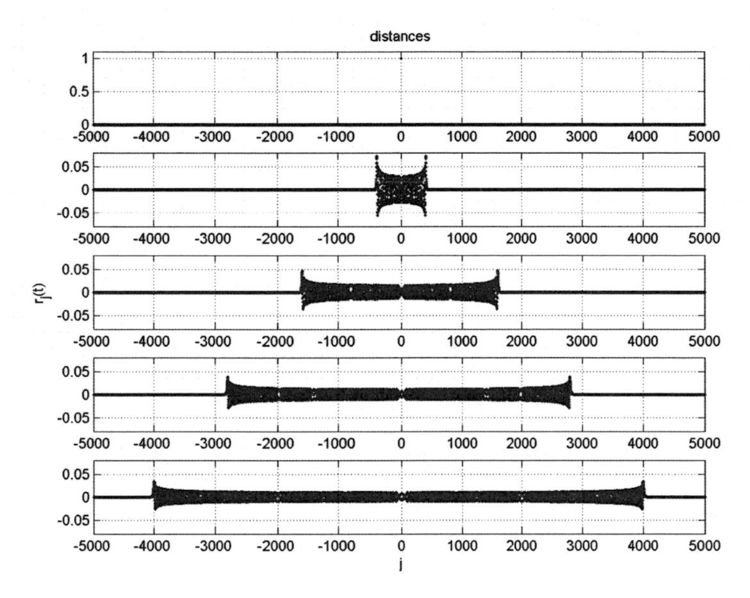

Figure 1.1: Solutions for different times for the linearized FPU $\ddot{x} = x_{j+1} - 2x_j + x_{j-1}$ with δ-initial conditions.

of the linearized system which is represented by oscillatory integrals. A careful study of the linear FPU equation was presented in [Fri03], which in particular highlights synchronization phenomena in compact domains. The dispersive decay is also discussed in detail in the upcoming Chapters 2 and 3.

A major property of the nonlinear system (1.2) is the existence of coherent structures relying on the interaction of dispersion and the nonlinearity. These solitary waves were first observed numerically by [FPU55] and later explained by formally deriving the completely integrable Korteweg-de Vries (KdV) equation in the long wave limit, cf. [ZK65]. In [FW94, FP99] the existence of solitary waves for generalized FPU systems is rigorously proven under additional global conditions on the interaction potentials V. Such waves satisfy $z_j(t) = Z(j-ct)$, where $z_j = (x_{j+1}-x_j, \dot{x}_j)$, for a fixed profile $Z : \mathbb{R} \to \mathbb{R}^2$ and a given wave speed c. In particular, [FW94] provides for the case $1 < \beta < 5$ the existence of solitary waves with arbitrarily small energy, i.e. $\|(\mathbf{x}^{\delta}_{\text{soli}}, \dot{\mathbf{x}}^{\delta}_{\text{soli}})\|_{\ell^2} = \delta \in (0, \delta_0)$.

Contrary to these results one might address the question of asymptotic stability. That is, asking when solitary waves do not occur but solutions decay in time due to a domination of the dispersive effects. In fact, the dispersive stability carries over to the nonlinear system as long as the nonlinearity is weak. In [GHM06, SK05] the authors consider a more general class of systems including (1.2) and prove asymptotic decay rates in ℓ^p if $\beta > 5$ and initial conditions are small and localized, i.e. small in ℓ^p with $p < 2$. In Chapter 2 these results are improved in a twofold manner: first we reduce the possible order of nonlinearity to the regime $\beta > 4$, and second we establish the better (and sharp) ℓ^p-decay rate $\alpha_p = (p-1)/(3p)$ in time. This result crucially relies on sharp estimates of the linearized system also proved in Chapter 2. For each $p \in [2,4) \cup (4,\infty]$ there exists a constant C_p such that

$$\|\mathrm{e}^{\mathcal{L}t}\mathbf{z}_0\|_{\ell^p} \leq \frac{C_p}{(1+t)^{\alpha_p}} \|\mathbf{z}_0\|_{\ell^1} \quad \text{for } t > 0$$

where the decay rates are given by

$$\alpha_p = \begin{cases} \frac{p-2}{2p} & \text{for } p \in [2,4) \\ \frac{p-1}{3p} & \text{for } p \in (4,\infty] \end{cases} .$$

Here $\mathcal{L}$ is the linear operator introduced in (1.3). The case $p = 4$ is excluded since the estimate holds only with a logarithmic correction.

The linear theory presented in Chapter 3 is in a sense an immediate continuation of that in Chapter 2. We derive an asymptotic expansion for the Green's function matrix that holds uniformly in space. Refining the ℓ^p-estimates this also resolves the microscopic oscillations. As a consequence we find that the ℓ^p-bounds are indeed sharp. To be more precise, we prove that there exists a constant $C(\omega)$ such that for all $t > 0$ the Green's function matrix $G_j(t) \in \mathbb{R}^{2\times 2}$ for $\dot{\mathbf{z}} = \mathcal{L}\mathbf{z}$ satisfies the estimate

$$|G_j(t) - \mathcal{G}^{\text{expan}}(t, j/t)| \leq C(\omega)/t \quad \text{for all } j \in \mathbb{Z} \text{ and all } t > 0.$$

The function $\mathcal{G}^{\text{expan}}(t,c)$ is determined explicitly. Figure 1.2 illustrates the two components of $\mathcal{G}^{\text{expan}}(t,c)$ caused by non-degenerated group velocities and degenerated ones characterized by $c = c_{\text{gr}}(\theta)$ and $\omega''(\theta)$. In the latter case the wave front matches to the plot of an Airy function. Apart from the fact that this result completes the picture of the behavior of the linearized system our hope is that the specific form of $\mathcal{G}^{\text{expan}}(t,c)$ allows for an improvement of the nonlinear stability results.

Now, returning to solitary waves, note that the stability result for the nonlinear FPU outlined above implies that for $\beta > 4$ these solution cannot be small in ℓ^1. In [FP99] the case $\beta = 2$ is investigated, and it is shown that there exist solitary waves traveling with velocity $c = c_s + \mathcal{O}(\varepsilon^2)$, where $c_s = |\omega'(0)|$ is the finite sonic wave speed of the linearized system. See also [McM02] and [FP02, FP04a, FP04b] in this context. The constructions there can be generalized to our case to provide small-energy solitary waves associated with the generalized KdV limit. Moreover, in [SW00] it was shown that solutions of the form $r_j^\varepsilon(t) = \varepsilon^{2/(\beta-1)}R(\varepsilon^3 t, \varepsilon(j+\omega'(0)t))+$ h.o.t. exist, with $R : [0,T] \times \mathbb{R} \to \mathbb{R}$ satisfying the generalized KdV equation

$$\partial_\tau R + b_1 \partial_\eta^3 R + b_2 \partial_\eta V'(R) = 0 .$$

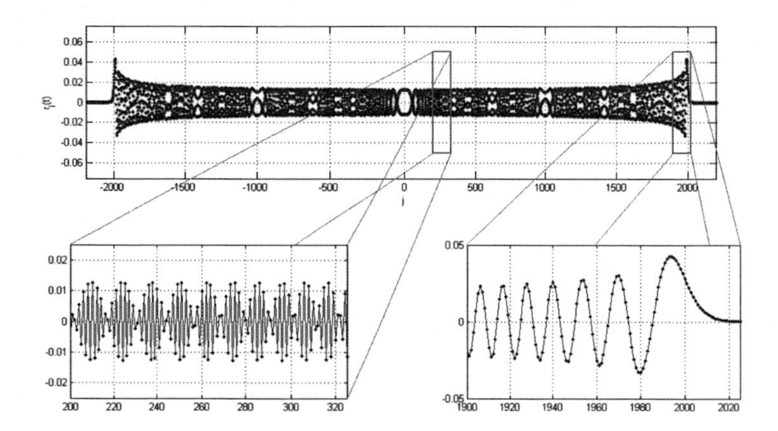

Figure 1.2: Green's function $G_{11}(t, j)$ at $t = 2000$ for the linear FPU chain. Lower left: periodic wave trains for non-degenerate group velocities. Lower right: Airy-type behavior at the degenerate front.

This equation possesses solitary wave solutions with exponentially decaying tailes. In terms of the generalized FPU system these solutions satisfy $\|\mathbf{z}^{\varepsilon}_{\text{soli}}(t)\|_{\ell^1} \sim \varepsilon^{(3-\beta)/(\beta-1)}$. This shows that for $1 < \beta < 3$ there are solitary waves that are arbitrarily small in ℓ^1. We conclude that the dispersive decay result mentioned above can not be transferred to $\beta < 3$. However, the case $\beta \in [3, 4]$ remains unresolved.

The work presented in Chapter 4 aims to close this gap for $\beta \in [3, 4]$, at least to the extend of presenting formal calculations and numerical experiments. Coming from Chapter 2 and 3 this serves as an outlook on future work. We expect the stability result to be suboptimal. In particular it is an open question whether the approach to treat the nonlinearity as a small perturbation of the linear part is exhausted. To answer this question, we analyze the relation between the order of the nonlinearity β and the question whether wave fronts emerging by the evolution of localized initial conditions are stable and dominated by linear or nonlinear dynamics. Using formal calculations and numerical experiments we show that for $\beta = 3$ the wave fronts still decay $\sim t^{-1/3}$ and behave like the solution of a nonlinear ODE, namely the 2nd Painlevé equation $\psi''(\zeta) = \zeta\psi(\zeta) + 2\psi(\zeta)^3$, which in fact is Airy's equation with an additional nonlinear term. In that case, perturbation arguments will fail and a substantial nonlinear theory is necessary. On the other hand, in case $\beta = 4$ we still see linear behavior. For $\beta \in (3, 4)$ the wave fronts decay like $t^{-1/3}$ as expected and form a crossover from Airy- to Painlevé-like behavior. We hope these insights form the basis to actually improve the stability result.

1.2 Thermodynamics

As an introduction to the last Chapter 5 we return to (1.1), this time in the context of statistical physics. Now we assume $\mathcal{J}$ to be finite. Given the potential energy E_{pot} of a system, finite temperature thermodynamical properties of materials are obtained from canonical ensemble averages,

$$\langle A \rangle = \frac{\int_{\Omega^{\mathcal{J}}} A(\mathbf{x}) \exp(-\beta E(\mathbf{x}))\, \mathrm{d}\mathbf{x}}{\int_{\Omega^{\mathcal{J}}} \exp(-\beta E(\mathbf{x}))\, \mathrm{d}\mathbf{x}}, \tag{1.4}$$

where $\Omega \subset \mathbb{R}$ is the macroscopic domain where the positions x_j vary, A is the observable of interest, and $\beta = 1/(k_B T)$ is the inverse temperature. The denominator of (1.4) is denoted by Z and called the partition function. The major computational difficulty is of course the N-fold integrals, where N, the number of particles, is extremely large. See [CLS07] for a review on sampling methods to compute (1.4).

It is often the case that the observable A does not actually depend on the positions x_i of all atoms, but only on some of them. Think for instance of nanoindentation, where we are especially interested in the positions of the atoms below the indenter. The Quasi-continuum Method is commonly used to compute such ensemble averages, cf. [SMT$^+$99, MT02, DTMP05, Leg09]. The basic idea of this approach is to divide the considered set of atoms in two sets $X_r \cup X_c = \{x_0, x_1 \ldots, x_J\}$ which are referred to as representative and constrained atoms, respectively. The decomposition is chosen such that the energy is basically described by the configuration of the representative atoms X_r. That is, we have approximately

$$\int \mathrm{e}^{-\beta E_{\text{pot}}(X_r, X_c)}\, \mathrm{d}X_r\, \mathrm{d}X_c \approx \int \mathrm{e}^{-\beta E_{\text{pot}}^{CG}(X_r)}\, \mathrm{d}X_r \tag{1.5}$$

or

$$E_{\text{pot}}^{CG}(X_r) \approx -\frac{1}{\beta} \ln \int \mathrm{e}^{-\beta E_{\text{pot}}(X_r, X_c)}\, \mathrm{d}X_c\,,$$

where E_{pot}^{CG} is the coarse grained potential energy. The key issue is again the efficient computation or approximation of this quantity. Here, the standard approach is to approximate the interaction potentials V_k by a harmonic potential and interpolate linearly to estimate the positions $x_j \in X_c$. Then E_{pot}^{CG} can be calculated exactly.

Without such simplifying assumptions, the actual computation of E_{pot}^{CG} for practical problems seems impossible. The approach has proven to be efficient. It satisfactorily treats three-dimensional problems of large size. However, from a mathematical point of view, it is an open question how to evaluate the impact of the approximations above. The purpose of the work presented in Chapter 5 is to present an approach that allows for a quantitative assessment of the validity and limits of these approximations in simple cases and under specific assumptions. This approach is based on the thermodynamic limit and is exact if the number of eliminated atoms is infinite. I.e. it provides a good approximation in $\#X_r \ll \#X_c$.

To illustrate the main idea of the limit process, consider $X_c = \{x_1, \ldots, x_{J-1}\}$ and assume $x_0 = 0$ and the density $\nu := x_J/J$ is fixed. We let x_j vary on the whole line $\mathbb{R}$. Since E_{pot}^{CG} scales with the number of particles, the right quantity to consider is the free energy per particle,

$$\mathcal{F}_J(\beta, \nu) = -\frac{1}{\beta J} \ln \int_{\mathbb{R}^{J-1}} \mathrm{e}^{-\beta \sum_{j=1}^{J} V(x_j - x_{j-1})} \, \mathrm{d}x_1 \ldots \mathrm{d}x_{J-1} \, . \tag{1.6}$$

First we consider pure nearest neighbour interaction with $V(r) = (r-1)^2$. Rewriting the potential energy in the form

$$\sum_{j=1}^{J} (x_j - x_{j-1} - 1)^2 = \tfrac{1}{J}(x_J - x_0 - aJ)^2 + \sum_{j=1}^{J-1} c_j (x_j - \xi_j)^2$$

with $c_j = (j+1)/j$ and $\xi_j = (j\, x_{j+1} + x_0)/(j+1)$ we find

$$\int_{\mathbb{R}^{J-1}} \mathrm{e}^{-\beta \sum_{j=1}^{J} V(x_j - x_{j-1})} \, \mathrm{d}x_1 \ldots \mathrm{d}x_{J-1} = \mathrm{e}^{-\beta(\nu - a)^2} \frac{1}{\sqrt{N}} \left(\frac{\pi}{\beta}\right)^{(N-1)/2}$$

Here we used in the translation invariance of the integral with respect to the integration variable the facts that $\prod_{j=1}^{J-1} c_j = J$ and $\int_{\mathbb{R}} \mathrm{e}^{-z^2} \mathrm{d}z = \sqrt{\pi}$. By inserting the last expression in (1.6) the free energy becomes

$$\begin{aligned} \mathcal{F}_J^{ha}(\nu, \beta) &= (\nu - a)^2 + \frac{1}{2\beta J} \ln J - \frac{J-1}{2\beta J} \ln \frac{\pi}{\beta} \\ &\to (\nu - a)^2 + \frac{1}{2\beta} \ln \frac{\beta}{\pi} \quad \text{as } J \to \infty \, . \end{aligned}$$

Note that this result is meaningful, for instance we obtain Hooke's law, $p = -\frac{\partial \mathcal{F}^{ha}}{\partial \nu} = -2(\nu - a)$.

In case of a general NN interaction we introduce a new variable z_j via $x_j = z_j + ja$ we rewrite the problem to get a more appropriate form to go to the limit. We choose a to be the mean

$$a := \int_{\mathbb{R}} y \, \frac{\mathrm{e}^{-\beta V(y)}}{Z} \, \mathrm{d}y \, , \qquad Z = \int_{\mathbb{R}} \mathrm{e}^{-\beta v(y)} \, \mathrm{d}y$$

and write $\mu(y) = \mathrm{e}^{-\beta V(y+a)}/Z$. This leads to

$$\mathcal{F}_J(\beta, \nu) = -\frac{1}{\beta J} \ln \left[\mu * \cdots * \mu\big((\nu - a)(J+1)\big)\right] - \frac{J+1}{\beta J} \ln Z \, ,$$

where the convolution is defined by $\mu * \mu(y) = \int_{\mathbb{R}} \mu(\rho)\mu(y - \rho) \, \mathrm{d}\rho$. Thus, the asymptotic behavior of the integral (1.6) is determined by that of the N-fold convolution of μ for arguments of order $\mathcal{O}(J)$.

The structure of the problem now gives rise to apply tools from probability theory. But note first that the central limit theorem fails: Given a sequence

$\{Y_i\}_{i\in\mathbb{N}}$ of i.i.d. random variables, then the Law of large numbers basically states that the density ρ_N of $S_N = \frac{1}{N}\sum_{i=1}^{N} Y_i$ converges weakly to the unit point measure concentrated at $\bar{y} = \mathbb{E}[Y_1]$. The Central limit theorem quantifies the probability that S_N differs from its mean $\bar{y}$ by an amount of order $\frac{1}{\sqrt{N}}$. But the large deviation property deals with events where S_N differs from $\bar{y}$ by an amount of $\mathcal{O}(1)$. More precisely, the law of large numbers states that for any Borel set A with $\bar{y} \notin \bar{A}$, $\rho_N(A) \to 0$ for $N \to \infty$. If the large deviation property holds for the sequence considered, then this convergence is of exponential rate which depends on the set A. Roughly speaking, a sequence $\{Y_i\}_{i\in\mathbb{N}}$ of i.i.d. random variables with density ρ has the *large deviation property* if there exists a sequence a_N, with $a_N \sim N$ as $N \to \infty$ such that for the density ρ_N of $S_N = \frac{1}{N}\sum_{i=1}^{N} Y_i$ holds

$$\rho_N(s) = \mathrm{e}^{-a_N \mathcal{I}(s) + r_N} \qquad \text{where} \quad r_N = o(N) \ \text{ as } \ N \to \infty\,. \tag{1.7}$$

A detailed definition is given in [Ell85a, Ell85b] and [DZ93], respectively.

In Chapter 5 we show that the theory of large deviations can indeed be applied to calculate (1.6). Based on this, a computational strategy to approximate ensemble averages is developed. Furthermore, computational results are collected.

Chapter 2

Dispersive stability of infinite-dimensional Hamiltonian systems on lattices

A sightly shortened version of the following paper by A. MIELKE and C. PATZ is published in *Applicable Analysis, Volume 89, Issue 9, 2010, p. 1493-1512* and also available as *WIAS Preprint No. 1448, Berlin 2009.*

The authors derive dispersive stability results for oscillator chains like the FPU chain or the discrete Klein-Gordon chain. If the nonlinearity is of degree higher than 4, then small localized initial data decay like in the linear case. For this, we provide sharp decay estimates for the linearized problem using oscillatory integrals and avoiding the non-optimal interpolation between different ℓ^p spaces.

Dispersive stability of infinite dimensional Hamiltonian systems on lattices*

Alexander Mielke[1,2], Carsten Patz[2]

[1] Weierstraß-Institut für Angewandte Analysis und Stochastik
Mohrenstraße 39, 10117 Berlin, Germany

[2] Humboldt-Universität zu Berlin, Institut für Mathematik
Rudower Chaussee 25, 12489 Berlin-Adlershof, Germany

5 October 2009. Revised 17 November 2009

Abstract

We derive dispersive stability results for oscillator chains like the FPU chain or the discrete Klein-Gordon chain. If the nonlinearity is of degree higher than 4, then small localized initial data decay like in the linear case. For this, we provide sharp decay estimates for the linearized problem using oscillatory integrals and avoiding the nonoptimal interpolation between different ℓ^p spaces.

Contents

*The research was partially supported by the Deutsche Forschungsgemeinschaft under Mie 459/4-3.

1 Introduction

The phenomenon of dispersive stability is well-studied for partial differential equations. Usually one considers a Hamiltonian system where energy conservation prevents strict spectral stability giving rise to exponential decay. Typically the behavior of small solutions is such that the energy norm is bounded from above and below by constants while the L^∞ norm decays with an algebraic rate of the type $(1+t)^{-\alpha}$. This rate is generated from the fact that initially localized solutions are dispersed by the different group velocities associated with the different wave numbers θ. The fundamental effects derive from the dispersion relation $\hat{\omega} = \omega(\theta)$ of the linear differential operator, where $\hat{\omega}$ is the frequency and $c(\theta) = \nabla_\theta \omega(\theta)$ the group velocity. The dispersion is now related to the fact that c still depends nontrivially on θ, i.e. the second derivative of ω should be nontrivial. We refer to [Seg68, Str74, Ree76, Str78] for results treating the sine-Gordon, the Klein-Gordon, the nonlinear Schrödinger, or the relativistic wave equations. Sometimes the theory is developed under the name scattering theory for small data. In [CW91] a recent improvement was made on the lowest order of nonlinearity for the generalized Korteweg-de Vries equation by a careful combination of sharp estimates for the linear part, obtained via deep harmonic analysis, and careful chain-rule estimates for fractional derivatives of the nonlinearity.

The same dispersive effects are to be expected in discrete systems, which are infinite ODEs on a lattice $\mathbb{Z}^d$. The difference is that the dispersion relation is now a periodic function in θ, i.e. ω is defined on the torus $\mathbb{T}^d := \mathbb{R}^d/_{(2\pi\mathbb{Z})^d}$. Thus, in contrast to PDEs, where ω is an algebraic function on $\mathbb{R}^d$, the dispersion relation has necessarily a richer degeneracy structure. As a result, the linear decay estimates for periodic lattices need a more careful analysis, and it is the aim of this work to establish a more general approach to this field.

To describe the work done so far and our contributions we start by highlighting the three major equations treated in this field, namely the Fermi-Pasta-Ulam chain (FPU), the Klein-Gordon chain (dKG) and the discrete nonlinear Schrödinger equation (dNLS):

$$\ddot{x}_j = V'(x_{j+1} - x_j) - V'(x_j - x_{j-1}), \qquad j \in \mathbb{Z}; \qquad \text{(FPU)}$$

$$\ddot{x}_j = x_{j+1} - 2x_j + x_{j-1} + W'(x_j), \qquad j \in \mathbb{Z}; \qquad \text{(dKG)}$$

$$\mathrm{i}\dot{u}_j = u_{j+1} - 2u_j + u_{j-1} + a|u_j|^{\beta-1}u_j, \qquad j \in \mathbb{Z}. \qquad \text{(dNLS)}$$

Here the potentials V and W are assumed to be such that $V'(r) = r + \mathcal{O}(|r|^\beta)$ and $W'(x) = x + \mathcal{O}(|x|^\beta)$. In general, $\beta > 1$ is used to measure the order of the nonlinearity.

A very careful study of the linear FPU equation was given in [Fri03], which highlights the synchronization phenomena in compact domains. In [Mie06] general multidimensional linear lattice systems were studied on the shorter hyperbolic scale, where energy transport along the rays dominates but dispersion is not yet seen. Discrete lattice systems as finite-difference approximations of wave equations are analyzed in [Zua05, IZ09], where the proper approximation of dispersion relations is an important point.

Dispersive stability results in the direction of this work are obtained in [SK05, GHM06]. The latter work provides the dispersive stability of FPU under the assumption that the nonlinearity satisfies $\beta > 5$. In this work, we will improve this result to the case $\beta > 4$. In [SK05] dKG and dNLS are studied analytically and numerically; we comment on the result of this paper below.

To describe our result we first restrict to FPU, which will be discussed in full detail in Section 3. There we will also treat a generalized FPU chain which allows for any finite number of interactions. Our main result will be that under a suitable stability and nonresonance condition we have dispersive stability if the nonlinearity is of order $\beta > 4$. In particular we will show that the decay of the solution of the nonlinear problem is the same as that of the linear one. The main point in the analysis is that we obtain an improved estimate for the dispersive decay of the linear semigroup. Writing FPU abstractly in the form

$$\dot{\mathbf{z}}(t) = \mathcal{L}\,\mathbf{z} + \mathcal{K}\mathcal{N}(\mathbf{z})$$

and using the Banach spaces $\ell^p = \ell^p(\mathbb{Z};\mathbb{R}^2)$ we find, for each $p \in [2,4) \cup (4,\infty]$, a constant C_p such that

$$\|\mathrm{e}^{\mathcal{L}t}\mathbf{z}_0\|_{\ell^p} \leq \frac{C_p}{(1+t)^{\alpha_p}}\,\|\mathbf{z}_0\|_{\ell^1}, \quad \|\mathrm{e}^{\mathcal{L}t}\mathcal{K}\,\mathbf{z}_0\|_{\ell^p} \leq \frac{C_p}{(1+t)^{\tilde{\alpha}_p}}\,\|\mathbf{z}_0\|_{\ell^1} \quad \text{for } t>0, \tag{1.1}$$

where the decay rates are given by

$$\alpha_p = \begin{cases} \dfrac{p-2}{2p} & \text{for } p \in [2,4), \\ \dfrac{p-1}{3p} & \text{for } p \in (4,\infty], \end{cases} \qquad \text{and} \qquad \tilde{\alpha}_p = \frac{p-2}{2p}\,.$$

The operator $\mathcal{K}$ arises from the difference structure of the right-hand side in FPU. The case $p = 4$ is excluded in (1.1), since the first estimate holds only with a logarithmic correction, see (3.9b).

The key observation is that the decay rates for $p \in (2,\infty)$ are strictly better than the ones obtained by interpolating the decay $\alpha_2 = 0$ and $\alpha_\infty = 1/3$, which would lead to $\hat{\alpha}_p = (p-2)/(3p) < \alpha_p$. The main work of Section 3 will be devoted to establish the decay estimates (1.1), which are obtained by analyzing the dispersion relation and estimating the resulting oscillatory integrals. The nonlinear stability result is then obtained using standard arguments, which we have collected in an abstract setting in Section 2. We emphasize that all nonlinear decay estimates are of the form that the nonlinear decay is exactly of the order as the linear decay, which is also found numerically, see Figure 1.1. We also show that our decay rates are optimal in the sense, that the dispersive decay of the nonlinear system cannot be better than for the linear system.

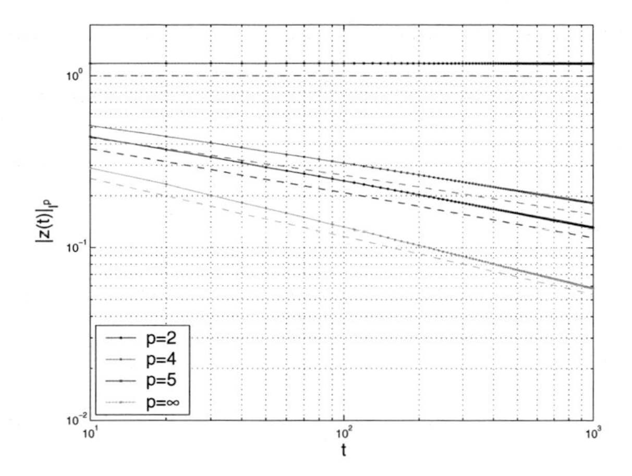

Figure 1.1: Double-logarithmic plot of ℓ^p norms of the solution to the linear FPU (—) and the nonlinear FPU with $V(r) = r + |r|^4$ (- -) as function of t.

In Section 4 we will discuss the usage of our method in more general settings such as dKG and dNLS. In particular, we compare our results for dKG with those obtained in [SK05]. There, for $\beta > 5$ dispersive decay in ℓ^p was proved with the rate $\hat{\alpha}_p = (p-2)/(3p)$, while numerically the values 0.226, 0.267, and 0.292 were obtained for $p = 4$, 5, and 6, respectively. We improve the results in a twofold manner: first we reduce the possible order of nonlinearity to the regime $\beta > 4$, and second we establish the better (and sharp) decay rate $\alpha_p = (p-1)/(3p)$ matching much better with the numerical values.

We conclude with remarking that there is a rich literature on persistent localized solutions in lattices, such as modulated pulses, solitons, and breathers, see e.g. [FW94, FP99, IJ05, GHM06]. From this, it is possible to show that the generalized FPU admits families of solitary waves of KdV type, which for $\beta < 5$ may have arbitrary small energy. However, these solitary waves are the broader the smaller the amplitude is. For the case $\beta < 3$ it follows that dispersive stability cannot hold, see Remark 3.3. It remains open what happens in the case $\beta \in [3, 4]$.

2 General stability result

In this section we present a general method to prove dispersive stability for nonlinear systems. The method, which is based on weak decay estimates of its linearization, is classical and was established for dispersive stability in PDE theory, see for instance [Seg68] and [Str74]. See also [MSU01] for a survey in the related theory of diffusive stability in parabolic systems. In the context of lattice models the authors of [GHM06] illustrate the ideas in an abstract setting and in [SK05] these arguments are applied to dKG systems and dNLS equations.

To emphasize the general structure we use again an abstract setting in general Banach spaces, which will be specialized to the spaces $\ell^p(\mathbb{Z}^d, \mathbb{R}^n)$ in the following sections. The general aim is to establish conditions that guarantee that the nonlinear system still has the same dispersive decay as the linear one. This will be our first result. In the second result we even go further by showing that the decay of the difference between the solution of the linear systems and the solution of the nonlinear systems is faster than the linear decay.

We start with the general system on a Banach space Z given in the form

$$\dot{\mathbf{z}} = \mathcal{L}\,\mathbf{z} + \mathcal{K}\mathcal{N}(\mathbf{z})\,, \tag{2.1}$$

where $\mathcal{L}$, $\mathcal{K}$ linear and bounded and $\mathcal{N}$ is a nonlinear operator. The operator $\mathcal{L} : Z \to Z$ generates a bounded semi-group $(\mathrm{e}^{\mathcal{L}t})_{t\geq 0}$, that is there exists a $C_{\mathcal{L}} > 0$ with $\|\mathrm{e}^{\mathcal{L}t}\mathbf{z}\|_Z \leq C_{\mathcal{L}}\|\mathbf{z}\|_Z$ for all $t \geq 0$ and $\mathbf{z} \in Z$. Typically the space Z is chosen such that the solution $\mathbf{z} = \mathbf{0}$ is a stable solution of (2.1), i.e.

$$\exists\, C_E > 0 \quad \forall\,\text{sln.}\ \mathbf{z}(t)\ \text{with}\ \|\mathbf{z}^0\|_Z \leq \varepsilon \quad \forall\, t > 0: \quad \|\mathbf{z}(t)\|_Z \leq C_E\|\mathbf{z}^0\|_Z. \tag{2.2}$$

This condition is in particular satisfied if the system is Hamiltonian and the energy functional serves as a Liapunov function. That is, if the energy is bounded from above and below.

However, for proving dispersive stability we need to choose different spaces and do not rely on (2.2). We consider a scale of Banach spaces $Z_0 \subset Z \subset X$ and a space $Z_{\mathcal{N}} \subset Z$ where the embeddings are assumed to be continuous. The space X is used for the estimation of the solutions, Z_0 is taken for the initial conditions, and $Z_{\mathcal{N}}$ measures the nonlinearity. We assume that positive constants $C_1, C_2, C_3, \alpha, \gamma$, and $\beta > 1$ exist such that the following estimates hold for all $\mathbf{z}$ and all $t \geq 0$:

$$\|\mathrm{e}^{\mathcal{L}t}\mathbf{z}\|_X \leq \frac{C_1}{(1+t)^\alpha}\|\mathbf{z}\|_{Z_0}, \tag{2.3a}$$

$$\|\mathrm{e}^{\mathcal{L}t}\mathcal{K}\,\mathbf{z}\|_X \leq \frac{C_2}{(1+t)^\gamma}\|\mathbf{z}\|_{Z_{\mathcal{N}}}, \tag{2.3b}$$

$$\|\,\mathcal{N}(\mathbf{z})\|_{Z_{\mathcal{N}}} \leq C_3\|\mathbf{z}\|_X^\beta. \tag{2.3c}$$

The following result is the first simple decay estimate, which we state for reasons of clarity. It is in fact a special case of the more involved result given below. Hence we do not provide an independent proof.

Theorem 2.1:
Let the conditions (2.3) *hold with* $\min\{\gamma, \alpha\beta, \alpha\beta{+}\gamma{-}1\} \geq \alpha$ *and* $\gamma \neq 1 \neq \beta\alpha$. *Then, there exist positive constants* C *and* ε *such that for each* $\mathbf{z}_0 \in Z_0$ *with* $\|\mathbf{z}_0\|_{Z_0} \leq \varepsilon$ *the unique solution* $\mathbf{z}$ *of* (2.1) *with* $\mathbf{z}(0) = \mathbf{z}_0$ *satisfies*

$$\|\mathbf{z}(t)\|_X \leq \frac{C}{(1+t)^\alpha}\|\mathbf{z}(0)\|_{Z_0} \quad \textit{for}\ t \geq 0.$$

This and the following result rely on the following lemma that is used to estimate the convolution integral occurring in the variation-of-constants formula. The lower bound in the following result is only given to indicate that the provided exponent γ is optimal.

Lemma 2.2:
For constants $\alpha_1, \alpha_2 \in [0,1) \cup (1,\infty)$ there exists a constant $C > 0$ such that

$$\frac{t}{C(1+t)^{\gamma+1}} \le \int_0^t \frac{1}{(1+t-s)^{\alpha_1}} \frac{1}{(1+s)^{\alpha_2}} \,\mathrm{d}s \le \frac{C}{(1+t)^{\gamma}} \quad \text{for all } t > 0, \tag{2.4}$$

where $\gamma = \min\{\alpha_1, \alpha_2, \alpha_1{+}\alpha_2{-}1\}$.

Proof. To obtain the estimate we split the integral into the two domains $[0, t/2]$ and $[t/2, t]$. In the first interval we estimate $(1+t)/2 \le 1+t-s \le 1+t$ and obtain

$$\frac{1}{(1+t)^{\alpha_1}} M_2(t/2) \le \int_0^{t/2} \frac{1}{(1+t-s)^{\alpha_1}} \frac{1}{(1+s)^{\alpha_2}} \,\mathrm{d}s \le \frac{2^{\alpha_1}}{(1+t)^{\alpha_1}} M_2(t/2)$$

where $M_2(r) = \int_0^r (1+s)^{-\alpha_2} \,\mathrm{d}s$. Evaluating the integral M_2 explicitly, we find a decay estimate with exponent $\gamma_2 = \min\{\alpha_1, \alpha_1{+}\alpha_2{-}1\}$. Treating the interval $[t/2, t]$ similarly, the assertion follows by taking $\gamma = \min\{\gamma_1, \gamma_2\}$. $\square$

The following result gives a refinement of the above result. It is based on an additional Banach space V which satisfies $Z_0 \subset V \subset X$ with continuous embeddings. It will play the role of an intermediate space in which we have already some information, namely

$$\exists C, \beta_1, \beta_2 > 0 \text{ with } \beta_1{+}\beta_2 > 1 \; \forall \mathbf{z} \in Z: \quad \|\mathcal{N}(\mathbf{z})\|_{Z_\mathcal{N}} \le C \|\mathbf{z}\|_V^{\beta_1} \|\mathbf{z}\|_X^{\beta_2}. \tag{2.5}$$

Such estimates occur naturally by interpolation, see (3.5).

Theorem 2.3:
Let the system (2.1) satisfy (2.3) and (2.5). Assume further that there exist positive δ, C_V, ν such that for all $\mathbf{z}_0 \in Z_0$ with $\|\mathbf{z}_0\|_{Z_0} \le \delta$ the unique solution $\mathbf{z}$ of (2.1) satisfies the estimate

$$\|\mathbf{z}(t)\|_V \le \frac{C_V}{(1+t)^{\nu}} \|\mathbf{z}(0)\|_{Z_0} \quad \text{for all } t \ge 0. \tag{2.6}$$

Let $\rho = \min\{\gamma, \beta_1\nu{+}\beta_2\alpha, \gamma{+}\beta_1\nu{+}\beta_2\alpha{-}1\}$ and assume $\rho \ge \alpha$, $\beta_1\nu{+}\beta_2\alpha \ne 1 \ne \gamma$, then there exist positive ε and C_X such that for $\|\mathbf{z}(0)\|_{Z_0} \le \varepsilon$ the solutions satisfy

$$\left.\begin{aligned} \|\mathbf{z}(t)\|_X &\le \frac{C_X}{(1+t)^{\alpha}} \|\mathbf{z}(0)\|_{Z_0} \quad \text{and} \\ \|\mathbf{z}(t) - \mathrm{e}^{\mathcal{L}t}\mathbf{z}(0)\|_X &\le \frac{C_X}{(1+t)^{\rho}} \|\mathbf{z}(0)\|_{Z_0}^{\beta_1+\beta_2} \end{aligned}\right\} \quad \text{for all } t \ge 0. \tag{2.7}$$

Proof. We give the proof in such a way that the case $\beta_1 = 0$ is included, which provides the proof of Theorem 2.1. Then, (2.5) reduces to (2.3c).

We use the variations-of-constants formula

$$\mathbf{z}(t) = \mathrm{e}^{\mathcal{L}t}\mathbf{z}(0) + \int_0^t \mathrm{e}^{\mathcal{L}(t-s)}\,\mathcal{K}\mathcal{N}(\mathbf{z}(s))\,\mathrm{d}s$$

and estimate the solution in the space X. Using the assumptions we obtain

$$\|\mathbf{z}(t)\|_X \leq \frac{C_1}{(1+t)^\alpha}\|\mathbf{z}(0)\|_{Z_0} + \int_0^t \frac{C_2}{(1+t-s)^\gamma}\frac{C(C_V\|\mathbf{z}(0)\|_{Z_0})^{\beta_1}}{(1+s)^{\nu\beta_1}}\|\mathbf{z}(s)\|_X^{\beta_2}\,\mathrm{d}s.$$

Assuming $\zeta = \|\mathbf{z}(0)\|_{Z_0} \leq \delta$ and introducing $R(t) = \max_{s\in[0,t]}(1+s)^\alpha\|\mathbf{z}(s)\|_X$ and $\mu = \beta_1\nu + \beta_2\alpha$ we find the estimate

$$R(t) \leq C_1\zeta + (1+t)^\alpha \int_0^t \frac{1}{(1+t-s)^\gamma}\frac{1}{(1+s)^\mu}\,\mathrm{d}s\, C_*\zeta^{\beta_1}R(t)^{\beta_2}.$$

Employing Lemma 2.2 we have derived the estimate $R(t) \leq C_1\zeta + C^*\zeta^{\beta_1}R(t)^{\beta_2}$. It is now easy to find $\varepsilon > 0$ such that for $\zeta \leq \varepsilon$ we have $R(t) \leq 2C_1\zeta$, which is the first inequality in (2.7).

Reconsidering the variations-of-constants formula once again gives

$$\|\mathbf{z}(t) - \mathrm{e}^{\mathcal{L}t}\mathbf{z}(0)\|_X \leq \int_0^t \frac{1}{(1+t-s)^\gamma}\frac{1}{(1+s)^\mu}\,\mathrm{d}s\, C_*\zeta^{\beta_1}R(t)^{\beta_2},$$

and the second estimate in (2.7) follows by employing Lemma 2.2 and the previous estimate for $R(t)$. □

3 Dispersive decay for generalized FPU systems

We now apply the general result presented in section 2 to Hamiltonian systems on a one-dimensional lattice, also called oscillator chain. Here, we only discuss a generalization of the celebrated Fermi-Past-Ulam chain in detail, while in section 4 we outline how to treat discrete Klein-Gordon systems and nonlinear Schrödinger equations.

3.1 The generalized FPU system

We consider an infinite number of equal particles with unit mass and interacting with a finite number K of neighbors via potentials $V_1, \dots, V_K$. According to Newton's law the equations of motion are

$$\ddot{x}_j = \sum_{1\leq k\leq K} \left(V_k'(x_{j+k} - x_j) - V_k'(x_j - x_{j-k})\right), \qquad j \in \mathbb{Z}. \tag{3.1}$$

Here $x_j \in \mathbb{R}$ denotes the displacements. We write $\mathbf{x} := (x_j)_{j\in\mathbb{Z}}$. For the time being we only assume that $V_k'(r) = a_k r + V_{\mathrm{nl},k}'(r)$, $V_{\mathrm{nl},k}'(r) = \mathcal{O}(|r|^\beta)_{|r|\to 0}$ with $\beta > 1$. System (3.1) is Hamiltonian, i.e. $(\dot{\mathbf{x}}, \dot{\mathbf{p}})^T = \mathcal{J}_{\mathsf{can}}\, \mathrm{d}\,\mathcal{H}_{\mathrm{x}}(\mathbf{x}, \mathbf{p})$ with momentum $\mathbf{p} := \dot{\mathbf{x}}$, $\mathcal{J}_{\mathsf{can}}$ the Poisson tensor corresponding to the canonical symplectic structure defined by $\langle(\mathbf{x},\mathbf{p}), \mathcal{J}_{\mathsf{can}}(\tilde{\mathbf{x}}, \tilde{\mathbf{p}})\rangle_{\ell^2\oplus\ell^2} = \langle \mathbf{x}, \tilde{\mathbf{p}}\rangle_{\ell^2} - \langle \tilde{\mathbf{x}}, \mathbf{p}\rangle_{\ell^2}$ and Hamiltonian

$$\mathcal{H}_{\mathrm{x}}(\mathbf{x}, \mathbf{p}) = \sum_{j\in\mathbb{Z}} \left(\frac{1}{2} p_j^2 + \sum_{1\le k\le K} V_k(x_{j+k} - x_j) \right).$$

The dispersive decay is driven by the linearized system

$$\ddot{x}_j = \sum_{1\le k\le K} a_k \big(x_{j+k} - 2x_j + x_{j-k} \big).$$

The dispersion relation is obtained by looking for plane waves in the form $x_j(t) = \mathrm{e}^{\mathrm{i}(\theta j + \hat{\omega} t)}$. We find the relation

$$\hat{\omega}^2 = \Lambda(\theta) := \sum_{1\le k\le K} a_k 2\big(1 - \cos(k\theta)\big). \tag{3.2}$$

Obviously, we have $\Lambda(0) = 0$, which is a consequence of Galilean invariance. By periodicity, it suffices to take $\theta \in [-\pi, \pi]$ and by reflection symmetry we may take $\theta \in [0, \pi]$ only. Throughout, we make the following stability condition

$$\Lambda(\theta) > 0 \quad \text{for all } \theta \in (0, \pi], \tag{3.3}$$

which certainly holds if all a_k are positive, however more general cases are possible.

An essential feature of the considered model is its Galilean invariance, i.e for all $\xi, c \in \mathbb{R}$ the transformation $(\mathbf{x}, \mathbf{p}) \mapsto (x_j + \xi + ct, p_j + c)_{j\in\mathbb{Z}}$ leaves (3.1) invariant. Therefore it is convenient to use distances $\mathbf{r} := (\partial_+ - \mathbf{1})\mathbf{x} = (x_{j+1} - x_j)_{j\in\mathbb{Z}}$ as new variables instead of the displacements. Introducing $\mathbf{z} := (\mathbf{r}, \mathbf{p})^T$ the Hamiltonian turns into

$$\mathcal{H}_{\mathrm{r}}(\mathbf{z}) = \frac{1}{2}\langle \mathbf{z}, \mathcal{A}_{\mathrm{r}}\, \mathbf{z}\rangle_{\ell^2} + \mathcal{V}_{\mathrm{nl}}(\mathbf{z})$$

with

$$\langle \mathbf{z}, \mathcal{A}_{\mathrm{r}}\, \mathbf{z}\rangle_{\ell^2} = \sum_{j\in\mathbb{Z}} \left(p_j^2 + \sum_{1\le k\le K} a_k \Big| \sum_{0\le l<k} r_{j+l} \Big|^2 \right)$$

and

$$\mathcal{V}_{\mathrm{nl}}(\mathbf{z}) = \sum_{j\in\mathbb{Z}} \sum_{1\le k\le K} V_{\mathrm{nl},k}\Big(\sum_{0\le l<k} r_{j+l} \Big).$$

The transformed Hamiltonian system (3.1) reads as

$$\dot{\mathbf{z}} = \mathcal{J}_{\mathrm{r}}\, \mathrm{d}\mathcal{H}_{\mathrm{r}}(\mathbf{z}) = \mathcal{L}\,\mathbf{z} + \mathcal{J}_{\mathrm{r}}\, \mathcal{N}(\mathbf{z}) \tag{3.4a}$$

where $\mathcal{L} = \mathcal{J}_\mathrm{r}\,\mathcal{A}_\mathrm{r}$ with $\mathcal{J}_\mathrm{r}$, $\mathcal{A}_\mathrm{r}$, and $\mathcal{N}$ given by

$$\mathcal{J}_\mathrm{r} := \begin{pmatrix} \mathbf{0} & \partial_+ - \mathbf{1} \\ \mathbf{1} - \partial_- & \mathbf{0} \end{pmatrix}, \quad \mathcal{A}_\mathrm{r} := \begin{pmatrix} \sum_{|l|<K} \sum_{|l|<k\le K} (k-|l|) a_k \partial_l & \mathbf{0} \\ \mathbf{0} & \mathbf{1} \end{pmatrix}, \tag{3.4b}$$

$$\mathcal{N}(\mathbf{z}) := \mathrm{d}\mathcal{V}_\mathrm{nl}(\mathbf{z}) = \begin{pmatrix} \left(\sum_{1\le k\le K} \sum_{0\le m<k} V'_{k,\mathrm{nl}}\big(\sum_{0\le l+m<k} r_{j+l}\big)\right)_{j\in\mathbb{Z}} \\ \mathbf{0} \end{pmatrix}, \tag{3.4c}$$

where $(\partial_l \mathbf{z})_j = z_{j+l}$ and $\partial_\pm = \partial_{\pm 1}$. Clearly, $\mathcal{L}\,\mathbf{z} = \mathcal{J}_\mathrm{r}\,\mathcal{A}_\mathrm{r}\,\mathbf{z}$ gives the linear forces and $\mathcal{J}_\mathrm{r}\,\mathcal{N}(\mathbf{z})$ the nonlinear interaction forces. Here the operator $\mathcal{J}_\mathrm{r}$ refers to the push-forward of the Poisson tensor $\mathcal{J}_\mathrm{can}$, i.e. $\mathcal{J}_\mathrm{r} = \mathcal{T}\,\mathcal{J}_\mathrm{can}\,\mathcal{T}^*$ where $\mathcal{T}$ is the linear map defined by $(\mathbf{r},\mathbf{p})^T = \mathcal{T}(\mathbf{x},\mathbf{p})^T$. Note that now $\mathcal{J}_\mathrm{r}$ is a non-canonical Poisson structure.

3.2 Nonlinear dispersive stability

To study the nonlinear system we use the Banach spaces

$$\ell^p(\mathbb{Z}^d;\mathbb{R}^m) \quad \text{with norm } \|\mathbf{z}\|_{\ell^p} := \Big(\sum\nolimits_{J\in\mathbb{Z}^d} |z_J|^p\Big)^{1/p}$$

where $p \in [1,\infty]$. We frequently write ℓ^p to denote $\ell^p(\mathbb{Z}^d;\mathbb{R}^m)$, if the lattice $\mathbb{Z}^d$ and the space $\mathbb{R}^m$ are either irrelevant or clear from the context.

For $1 \le p_1 < p_2 \le \infty$ we have the continuous embedding $\ell^{p_1} \subset \ell^{p_2}$ with $\|\mathbf{z}\|_{\ell^{p_2}} \le \|\mathbf{z}\|_{\ell^{p_1}}$. An essential tool is the interpolation estimate

$$\|\mathbf{z}\|_{\ell^{p_\vartheta}} \le \|\mathbf{z}\|_{\ell^{p_0}}^{1-\vartheta} \|\mathbf{z}\|_{\ell^{p_1}}^{\vartheta}, \quad \text{where } \frac{1}{p_\vartheta} = \frac{1-\vartheta}{p_0} + \frac{\vartheta}{p_1}, \tag{3.5}$$

$p_0, p_1 \in [1,\infty]$ and $\vartheta \in [0,1]$. This is an easy consequence of Hölder's inequality and plays a crucial role in many estimates concerning dispersive decay. Moreover, we use Young's inequality for convolutions $a * b$ with $(a*b)_J = \sum_{I\in\mathbb{Z}^d} a_{J-I} b_I$. For $r,p,q \in [1,\infty]$ with $\frac{1}{p} + \frac{1}{q} = 1 + \frac{1}{r}$ we have

$$\|a*b\|_{\ell^r} \le \|a\|_{\ell^p} \|b\|_{\ell^q} \quad \text{for all } a \in \ell^p,\ b \in \ell^q. \tag{3.6}$$

To apply the general result of section 2 we first provide the a priori estimate (2.2). The theory in section 3.3 shows that (3.3) is equivalent to the existence of a constant $C \ge 1$ such that

$$\frac{1}{C}\|\mathbf{z}\|_{\ell^2}^2 \le \langle \mathbf{z}, \mathcal{A}_\mathrm{r}\,\mathbf{z}\rangle_{\ell^2} \le C\|\mathbf{z}\|_{\ell^2}^2 \quad \text{for all } \mathbf{z} \in \ell^2(\mathbb{Z};\mathbb{R}^2).$$

Using this it is easy to obtain the classical energy stability in $\ell^2(\mathbb{Z};\mathbb{R}^2)$: there are $C_2 > 0$ and $\varepsilon_0 > 0$ such that for all $\mathbf{z}_0 \in \ell^2$ with $\|\mathbf{z}_0\|_{\ell^2} \le \varepsilon_0$ the solution $\mathbf{z}$ of (3.4) with $\mathbf{z}(0) = \mathbf{z}_0$ exists globally in time and satisfies

$$\|\mathbf{z}(t)\|_{\ell^2} \le C\|\mathbf{z}(0)\|_{\ell^2} \quad \text{for all } t \in \mathbb{R}. \tag{3.7}$$

To state the linear decay result we define the relevant branch $\hat{\omega} = \omega(\theta)$ of the dispersion relation via

$$\omega(\theta) := \sqrt{\Lambda(\theta)} \geq 0.$$

With a slight abuse of notation we simply call ω the dispersion relation. Under the stability assumptions (3.3) we have $\omega \in \mathrm{C}^\infty([0,\pi])$ and we are able to define the set of critical wave numbers as

$$\Theta_{\mathrm{cr}} := \{\theta \in [0,\pi] \mid \omega''(\theta) = 0\}.$$

Since K in (3.2) is finite, Θ_{cr} is discrete and contains $\theta = 0$. Thus, we have $\Theta_{\mathrm{cr}} = \{\theta_0, \ldots, \theta_M\}$ with $\theta_0 = 0 < \theta_1 < \ldots < \theta_M \leq \pi$ for some $M \in \mathbb{N}$.

The following linear decay results will be proved in section 3.3.

Theorem 3.1:
Consider the group $(\mathrm{e}^{\mathcal{L}t})_{t\in\mathbb{R}}$ for $\mathcal{L} = \mathcal{J}_\mathrm{r}\mathcal{A}_\mathrm{r}$ defined in (3.4b). Assume that the dispersion relation ω satisfies (3.3) and the non-degeneracy condition

$$\omega'(0) > 0 \quad \text{and} \quad \forall\, \theta \in \Theta_{\mathrm{cr}} : \ \omega'''(\theta) \neq 0. \tag{3.8}$$

Then, for $p \in [2,4) \cup (4,\infty]$ there exists C_p such that, for all $t \geq 0$, we have

$$\|\mathrm{e}^{\mathcal{L}t}\|_{\ell^1,\ell^p} \leq \frac{C_p}{(1+t)^{\alpha_p}}, \quad \text{where} \quad \alpha_p = \begin{cases} \dfrac{p-2}{2p} & \text{for } p \in [2,4), \\ \dfrac{p-1}{3p} & \text{for } p \in (4,\infty]. \end{cases} \tag{3.9a}$$

In the case $p = 4$ there exists $C_4 > 0$ such that

$$\|\mathrm{e}^{\mathcal{L}t}\|_{\ell^1,\ell^4} \leq C_4 \Big(\frac{\log(2+t)}{1+t}\Big)^{1/4} \quad \text{for all } t > 0. \tag{3.9b}$$

If furthermore $\Theta_{\mathrm{cr}} = \{0\}$, then for $p \in [2,\infty]$ there exists $\tilde{C}_p$ such that

$$\|\mathrm{e}^{\mathcal{L}t}\,\mathcal{J}_\mathrm{r}\|_{\ell^1,\ell^p} \leq \frac{\tilde{C}_p}{(1+t)^{\tilde{\alpha}_p}} \quad \text{for all } t > 0, \quad \text{where } \tilde{\alpha}_p = \frac{p-2}{2p}. \tag{3.10}$$

The philosophy of the decay estimate is that oscillations with wave numbers θ travel along rays $j = c(\theta)t$, where the group velocity is given by $c(\theta) = \omega'(\theta)$. The decay along these rays is like $t^{-1/2}$ if $\omega''(\theta) \neq 0$ and like $t^{-1/3}$ if $\theta \in \Theta_{\mathrm{cr}}$. In Figure 3.1 we plot the dispersion relations ω and the associated solution $r_j(t)$ to display the influence of the critical wave numbers $\theta_j \in \Theta_{\mathrm{cr}}$. Thus, the decay like $t^{-1/3}$ in ℓ^∞ is easily obtained. However, for $\theta \approx \theta_n \in \Theta_{\mathrm{cr}}$ there is a cross-over between the two different decay rates, which needs to be estimated carefully to obtain the decay rate α_p for $p \in (2,\infty)$.

Because the operator $\mathcal{J}_\mathrm{r}$ is related to the difference operators $\partial_+ - 1$ and $1 - \partial_-$, it reduces the amplitudes of very long waves. Thus, in $\mathrm{e}^{\mathcal{L}t}\,\mathcal{J}_\mathrm{r}$ the bad decay associated with $\theta_0 = 0 \in \Theta_{\mathrm{cr}}$ is reduced but not for any other $\theta_n \in \Theta_{\mathrm{cr}}$. Hence, the

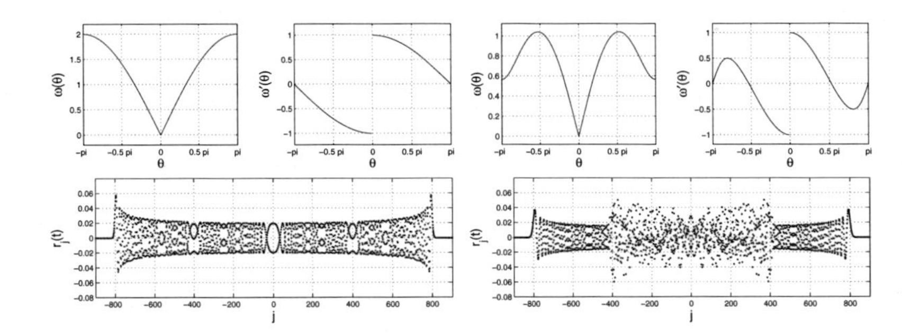

Figure 3.1: Dispersion relations and time evolutions. Left: classical FPU ($K = 1$: $a_1 = -1$). Right: generalized FPU ($K = 2$: $a_1 = 0.08$, $a_2 = 0.23$) with two wave fronts. The upper figures show $\omega(\theta)$ and $\omega'(\theta)$, respectively, and the lower figure shows $r_j(t)$ for $t = 800$ and initial condition $(r_j(0), \dot{x}_j(0)) = (\delta_{j,0}, 0)$.

last statement needs the requirement $\Theta_{\rm cr} = \{0\}$. In this connection it is interesting to mention that in case $\Theta_{\rm cr} = \{0\}$ the solutions of $\dot{\mathbf{z}} = \mathcal{L}\,\mathbf{z}$ globally decay like $t^{-1/2}$ if we restrict the initial conditions to a suitable subspace. Indeed, if we choose $\mathbf{z}^0 \in \mathcal{J}_{\rm r}\ell^1$ this follows from (3.10) and the fact that the operators $\mathcal{J}_{\rm r}$ and $\mathrm{e}^{\mathcal{L}t}$ commute.

The following decay result is a direct combination of the abstract results of section 2 and the above linear decay estimates.

Theorem 3.2:
Consider the generalized FPU system satisfying the linearized stability condition (3.3) *and the non-degeneracy condition* (3.8). *Assume that each potential V_k satisfies $V_k'(r) = a_k r + \mathcal{O}(|r|^\beta)$ for $\beta > 4$. Then, for each $p \in [2,4) \cup (4,\infty]$ there exist C_p and $\varepsilon > 0$ such that all solutions $\mathbf{z}$ of* (3.4) *with $\|\mathbf{z}(0)\|_{\ell^1} \leq \varepsilon$ satisfy the estimate*

$$\|\mathbf{z}(t)\|_{\ell^p} \leq \frac{C_p}{(1+t)^{\alpha_p}} \|\mathbf{z}(0)\|_{\ell^1} \quad \textit{for all } t \geq 0, \tag{3.11}$$

where the decay rate α_p is given in (3.9). *If additionally $\Theta_{\rm cr} = \{0\}$, then*

$$\|\mathbf{z}(t) - \mathrm{e}^{\mathcal{L}t}\mathbf{z}(0)\|_{\ell^p} \leq \frac{\tilde{C}_p}{(1+t)^{\tilde{\alpha_p}}} \|\mathbf{z}(0)\|_{\ell^1}^\beta \quad \textit{for all } t \geq 0, \tag{3.12}$$

where the decay rate $\tilde{\alpha_p}$ is given in (3.10).

We have omitted the case $p = 4$ to avoid a clumsy presentation. For $p = 4$ one can easily obtain algebraic decay for any $\alpha < 1/4$ by interpolation or a decay as in (3.9b), after generalizing the results in Section 2 to include logarithmic terms.

Proof. In a first step we apply Theorem 2.1 with $Z_0 = Z_{\mathcal{N}} = \ell^1 \subset X = \ell^{p_1}$ with $4 > p_1 > 2\beta/(\beta-2)$, where we used $\beta > 4$. Because of $\beta > p_1$ we have $\|\mathcal{N}(\mathbf{z})\|_{\ell^1} \leq C\|\mathbf{z}\|_{\ell^{p_1}}^{\beta}$. We estimate $\mathcal{K} = \mathcal{J}_{\mathrm{r}}$ by a constant and use $\alpha = \gamma = \alpha_{p_1}$. The choice of p_1 gives $\alpha < 1 < \alpha\beta$ and $\min\{\gamma, \alpha\beta, \alpha\beta+\gamma-1\} = \alpha$, which allows us to apply the theorem. We obtain positive C_{p_1} and ε_0 such that (3.11) holds for $p = p_1$. Since the result holds for $p = 2$ by the nonlinear stability estimate (3.7), the interpolation (3.5) shows that the result holds for $p \in [2, p_1]$. Since p_1 can be chosen as close to $p = 4$ as we like, estimate (3.11) is established for all $p \in [2, 4)$.

Next we consider $p \in (4, \beta]$ and see that Theorem 2.3 is applicable with $\nu = \beta_1 = 0$, $\alpha = \alpha_p < \gamma = \tilde{\alpha}_p$, and $Z_0 = Z_{\mathcal{N}} = \ell^1 \subset X = \ell^p$. Thus, (3.11) and (3.12) hold for $p \in (4, \beta]$.

Finally, we treat the case $p = \infty$ by choosing $p_2 \in (2, 4)$ with $p_2 \geq 12-2\beta < 4$. Using $\|\mathcal{N}(\mathbf{z})\|_{\ell^1} \leq C\|\mathbf{z}\|_{\ell^{p_2}}^{p_2}\|\mathbf{z}\|_{\ell^\infty}^{\beta-p_2}$ we are able to employ Theorem 2.3 with $Z_0 = Z_{\mathcal{N}} = \ell^1 \subset V = \ell^{p_2} \subset X = \ell^\infty$, and $\nu = \alpha_{p_2} < \alpha = 1/3 \leq \gamma$, where $\gamma = 1/3$ in the general case and $\gamma = 1/2$ if the additional condition $\Theta_{\mathrm{cr}} = \{0\}$ holds, see Theorem 3.1. Using $\beta_1 = p_2$ and $\beta_2 = \beta - p_2$ we find $\nu\beta_1 + \alpha\beta_2 > 1$. Hence $\rho = \gamma \geq \alpha$ and the desired estimate (3.11) follows for $p = \infty$. Again, the remaining range $p \in [\beta, \infty]$ follows from interpolation.

If the additional condition $\Theta_{\mathrm{cr}} = \{0\}$ holds, we can apply the last assertion in Theorem 2.3 and obtain (3.12). □

So far, we have only derived estimates for $Z_{\mathcal{N}} = \ell^1$. It is however straight forward to obtain results for $Z_{\mathcal{N}} = \ell^q$ for $q \in (1, 2)$, however the decay rates will be lower and one may need higher order of nonlinearity β. To see this, we simply note that the application of the operator $\mathrm{e}^{\mathcal{L}t}$ is in fact a convolution with a matrix-valued Green's function $\mathbf{G}(t) \in \ell^1(\mathbb{Z}; \mathbb{R}^{2\times 2})$, cf. (3.20). Hence, using Young's inequality (3.6) the operator norm $\|\mathrm{e}^{\mathcal{L}t}\|_{\ell^q,\ell^p}$ can be estimated by $\|\mathbf{G}\|_{\ell^s}$ where $1 + \frac{1}{p} = \frac{1}{s} + \frac{1}{q}$. In fact, the estimates stated above and proved below are obtained by estimating the ℓ^p norm of $\mathbf{G}(t)$.

We emphasize that for $q = 1$ the formula $\|\mathrm{e}^{\mathcal{L}t}\|_{\ell^1,\ell^p} = \|\mathbf{G}(t)\|_{\ell^p}$ holds, since the upper bound follows from Young's inequality and the lower bound is obtained by using the initial condition $\mathbf{z} = (\delta_j)_{j\in\mathbb{Z}}$. Our estimates for $\mathbf{G}_j(t)$ will be sharp enough to establish also lower bounds $\|\mathbf{G}(t)\|_{\ell^p} \geq c/(1+t)^{\alpha_p}$. Thus, we cannot hope for better estimates for the linear terms. In fact, using that the decay rates $\alpha_2 = 0$ and $\alpha_\infty = 1/3$ are optimal, it suffices to show that the decay rate α_4 cannot be better than $1/4$ (up to the logarithmic term). Then, for no $p \in (2, \infty)$ the decay rate can be better than α_p, because an interpolation would lead to a better decay rate for $p = 4$. Below we will show that estimate (3.9b) is indeed optimal.

Figure 3.2 displays numerically estimated decay rates, the exact curve α_p, and the curve $\hat{\alpha}_p = (p-2)/(3p)$, which is obtained by interpolation between $p = 2$ and $p = \infty$ and hence is not optimal. The numerical curves agrees well with α_p away from $p = 4$. This effect may be due to the logarithmic correction which spoils the convergence.

In the following remark we argue that the above dispersive decay cannot hold

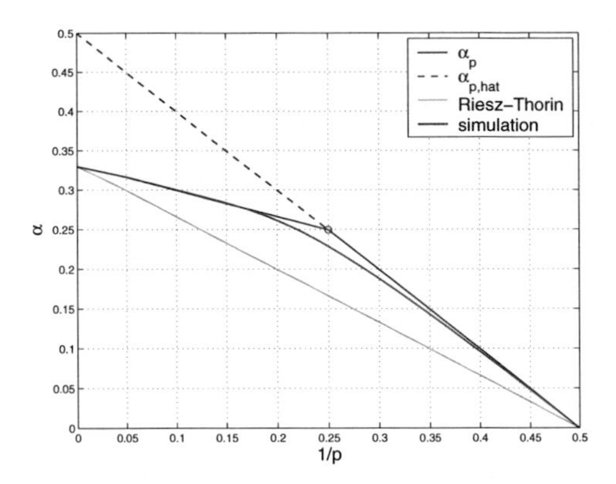

Figure 3.2: Exact decay rate α_p, interpolation rate $\hat{\alpha}_p$, ℓ^2-ℓ^∞ interpolation rate and numerically estimated rates as functions of $1/p$.

for $\beta < 3$, because of existence of solitary waves with arbitrary small ℓ^1 norm.

Remark 3.3 (Solitary waves):
From [FW94, FP99] the existence of solitary waves for generalized FPU systems can be deduced under additional global conditions on the interaction potentials V_k. Such waves satisfy $z_j(t) = Z(j-ct)$ for a fixed profile $Z : \mathbb{R} \to \mathbb{R}^2$ and a given wave speed c. In particular, [FW94] provides for the case $1 < \beta < 5$ the existence of solitary waves with arbitrarily small energy, i.e. $\|\mathbf{z}^\delta_{\mathrm{soli}}\|_{\ell^2} = \delta \in (0, \delta_0)$. Our stability result implies that for $\beta > 4$ these solution cannot be small in ℓ^1.

In [FP99] the case $\beta = 2$ is investigated, and it is shown that $c = \omega'(0) + \mathcal{O}(\varepsilon^2)$. The constructions there can be generalized to our case to provide small-energy solitary waves of associated with the generalized KdV limit. Moreover, in [SW00] it was shown that solutions of the form $r^\varepsilon_j(t) = \varepsilon^{2/(\beta-1)} R(\varepsilon^3 t, \varepsilon(j+\omega'(0)t)) + \text{h.o.t.}$ exist, with $R : [0,T] \times \mathbb{R} \to \mathbb{R}$ satisfying the generalized KdV equation

$$\partial_\tau R + b_1 \partial_\eta^3 R + b_2 \partial_\eta V'(R) = 0$$

where $V'(r) = r + \mathcal{O}(|r|^\beta)$. This equation possesses solitary wave solutions with exponentially decaying tales. In terms of the generalized FPU system these solutions satisfy $\|\mathbf{z}^\varepsilon_{\mathrm{soli}}(t)\|_{\ell^1} \sim \varepsilon^{(3-\beta)/(\beta-1)}$, which shows that for $1 < \beta < 3$ there are solitary waves that are arbitrarily small in ℓ^1. We conclude that the above dispersive decay result cannot hold for $\beta < 3$, while the case $\beta \in [3, 4]$ remains open.

3.3 ℓ^p-estimates for the linearized system

We consider the linearization of (3.4) in $\mathbf{z} = \mathbf{0}$, i.e. the case $\mathcal{N}(\mathbf{z}) \equiv 0$. To solve the system explicitly we use Fourier transform $\mathcal{F}: \ell^2(\mathbb{Z}, \mathbb{R}^2) \to L^2(\mathcal{S}^1, \mathbb{R}^2)$ defined by $\hat{z}(\theta) = \sum_{j\in\mathbb{Z}} z_j \mathrm{e}^{-\mathrm{i}j\theta}$. Then $\dot{\mathbf{z}} = \mathcal{J}_\mathrm{r} \mathcal{A}_\mathrm{r} \mathbf{z}$ turns into

$$\begin{pmatrix} \dot{\hat{\mathbf{r}}} \\ \dot{\hat{\mathbf{p}}} \end{pmatrix} = \begin{pmatrix} 0 & \mathrm{e}^{\mathrm{i}\theta} - 1 \\ 1 - \mathrm{e}^{-\mathrm{i}\theta} & 0 \end{pmatrix} \begin{pmatrix} \omega_\mathrm{r}^2(\theta) & 0 \\ 0 & 1 \end{pmatrix} \cdot \begin{pmatrix} \hat{\mathbf{r}} \\ \hat{\mathbf{p}} \end{pmatrix}, \tag{3.13}$$

where

$$\begin{aligned} \omega_\mathrm{r}^2(\theta) &= \sum_{|l|\le K-1} \sum_{|l|<k\le K} (k - |l|) a_k \mathrm{e}^{\mathrm{i}l\cdot\theta} \\ &= \sum_{0<k\le K} k a_k + 2 \sum_{0<l\le K-1} \Big(\sum_{l<k\le K} (k-l) a_k \Big) \cos(l\cdot\theta). \end{aligned} \tag{3.14}$$

Since (3.13) implies $\ddot{\hat{\mathbf{r}}} = 2(\cos\theta - 1)\omega_r(\theta)^2 \hat{\mathbf{r}}$ and since the previous subsection gives $\ddot{\hat{\mathbf{r}}} = -\Lambda(\theta)\hat{\mathbf{r}}$ we conclude

$$\omega(\theta) = 2\left|\sin\frac{\theta}{2}\right| \omega_\mathrm{r}(\theta). \tag{3.15}$$

Using the linear stability condition (3.3) and the non-degeneracy condition (3.8) we obtain

$$\exists\, c_\mathrm{r} > 0\ \forall \theta \in \mathcal{S}^1: \qquad \omega_\mathrm{r}(\theta) = \omega_\mathrm{r}(-\theta) \ge c_\mathrm{r}. \tag{3.16}$$

The fundamental matrix of the linear system (3.13) is

$$\hat{\mathbf{G}}_\mathrm{r}(\theta, t) = \begin{pmatrix} \cos(\omega(\theta)t) & \frac{\mathrm{e}^{\mathrm{i}\theta}-1}{\omega(\theta)} \sin(\omega(\theta)t) \\ \frac{-\omega(\theta)}{\mathrm{e}^{\mathrm{i}\theta}-1} \sin(\omega(\theta)t) & \cos(\omega(\theta)t) \end{pmatrix}. \tag{3.17}$$

The Green's function of our original problem is given by $\mathbf{G}(t) = \mathcal{F}^{-1}\hat{\mathbf{G}}_\mathrm{r}(\theta, t)$, that is $G_j(t) = \frac{1}{2\pi}\int_{\mathcal{S}^1} \hat{G}_\mathrm{r}(\theta, t) \mathrm{e}^{\mathrm{i}j\theta}\,\mathrm{d}\theta$ for $j \in \mathbb{Z}$. Thus the long time behavior of solutions to the linearized system is determined by oscillatory integrals. For instance, for the classical FPU, i.e. for $\omega_\mathrm{r}(\theta) \equiv 1$, the components of G_j turn into Bessel functions, see [Fri03]. Below we apply tools from asymptotic analysis to obtain upper bounds on the solutions of the linearized system. To do so it turns out that an alternative representation of the above Green's function is more convenient. Using the symmetry of ω_r we find that

$$G_j(t) = \frac{1}{2\pi}\int_0^\pi \begin{pmatrix} h_\pm(\theta, t, \frac{j}{t}) & \frac{1}{\omega_\mathrm{r}(\theta)} h_\pm\big(\theta, t, \frac{j+1/2}{t}\big) \\ \omega_\mathrm{r}(\theta) h_\pm\big(\theta, t, \frac{j-1/2}{t}\big) & h_\pm(\theta, t, \frac{j}{t}) \end{pmatrix} \mathrm{d}\theta \tag{3.18}$$

with $h_\pm(\theta, t, c) = \cos\big(t(\omega(\theta) + \theta c)\big) \pm \cos\big(t(\omega(\theta) - \theta c)\big)$.

The new variable $c \in \mathbb{R}$ roughly characterizes the rays $j = ct$ and is used to remind us to the group velocity $c(\theta) = \omega'(\theta)$.

Thus, we obtained the following representation formula for the solution of the linearized problem.

Lemma 3.4 (Explicit solution):
Given some initial conditions $\mathbf{z}^0 = (\mathbf{r}^0, \mathbf{p}^0)^T \in \ell^2(\mathbb{Z}, \mathbb{R}^2)$, *the unique solution of* $\dot{\mathbf{z}} = \mathcal{L}\,\mathbf{z}$ *with* $\mathcal{L} = \mathcal{J}_\mathrm{r}\,\mathcal{L}_\mathrm{r}$ *defined in* (3.4b) *is determined by*

$$\mathbf{z}(t) = \mathrm{e}^{\mathcal{L}\,t}\mathbf{z}^0 \tag{3.19}$$

where $(\mathrm{e}^{\mathcal{L}\,t})_{t\in\mathbb{R}}$ *is a differentiable group of bounded operators on* $\ell^2(\mathbb{Z}, \mathbb{R}^2)$ *defined by*

$$\left(\mathrm{e}^{\mathcal{L}\,t}\mathbf{z}\right)_j = \sum_{k\in\mathbb{Z}} G_k(t)\cdot z_{j-k} \qquad \text{for } j\in\mathbb{Z} \tag{3.20}$$

with $G_j(t)$ *defined in* (3.18).

The asymptotic behavior of (3.18) is determined by terms of the form

$$g(t,c) = \int_0^\pi \psi(\theta)\mathrm{e}^{\mathrm{i}t\phi(\theta,c)}\,\mathrm{d}\theta \quad \text{with } \phi(\theta,c) = (\omega(\theta)\pm c\theta) \tag{3.21}$$

with ω defined in (3.15) and $\psi(\theta)$ standing for 1, $1/\omega_\mathrm{r}(\theta)$ or $\omega_\mathrm{r}(\theta)$. In any case ψ is smooth on $[0,\pi]$.

The main result from asymptotic analysis we will use below is van der Corput's lemma, see e.g. [Ste93]. It states that if ϕ is smooth and $\left|\phi^{(k)}(\theta)\right| \geq \lambda > 0$ for $\theta\in(a,b)$ where either $k\geq 2$, or $k=1$ and ϕ' is monotonic, then

$$\left|\int_a^b \mathrm{e}^{\mathrm{i}t\phi(\theta)}\,\mathrm{d}\theta\right| \;\leq\; C_k\,(\lambda t)^{-\frac{1}{k}} \quad \text{with} \quad C_k = (5\cdot 2^{k-1}-2). \tag{3.22}$$

Note that C_k does neither depend on a and b nor on ϕ explicitly. Writing $F(\theta) = \int_a^\theta \mathrm{e}^{\mathrm{i}t\phi(\xi)}\,\mathrm{d}\xi$ and applying integration by parts to $\int_a^b \psi(\theta)F'(\theta)\mathrm{d}\theta$ we obtain

$$\left|\int_a^b \mathrm{e}^{\mathrm{i}t\phi(\theta)}\psi(\theta)\,\mathrm{d}\theta\right| \;\leq\; C_k\,(\lambda t)^{-\frac{1}{k}}\left(|\psi(b)| + \int_a^b |\psi'(\theta)|\,\mathrm{d}\theta\right).. \tag{3.23}$$

In the following lemmas we provide the decay estimates on $g(t,c)$ required to prove the sharp ℓ^p decay rate of the linear group $\mathrm{e}^{\mathcal{L}\,t}$. We use the notation

$$C_\psi := \max_{\theta\in[0,\pi]} |\psi(\theta)| + \int_0^\pi |\psi'(\theta)|\,\mathrm{d}\theta.$$

Since van der Corput's Lemma only demands assumptions on $\left|\phi^{(k)}(\theta)\right|$ the following considerations are indeed independent of the sign of ϕ in (3.21).

The first lemma provides a global upper bound on $g(t,c)$. Using the classical method of stationary phase, see [Won89], it is straight forward to check that the result is sharp.

Lemma 3.5 (Global bound):
Consider the oscillatory integral (3.21) *with dispersion relation* ω *satisfying* (3.8)

and $\psi \in W^{1,1}([0,\pi])$. Then there exists a constant $C_\omega > 0$ depending only on ω such that

$$\forall t \geq 0,\ c \in \mathbb{R}: \quad |g(t,c)| \leq \frac{C_\omega C_\psi}{(1+t)^{1/3}}. \tag{3.24}$$

Proof. Due to $\phi(\theta, c) = \omega''(\theta)$ the following considerations are uniform with respect to the group velocity c.

We write $U_\delta(\theta_m) = \{\theta \in [0,\pi] \mid |\theta - \theta_m| < \delta\}$. Due to the non-degeneracy condition (3.8) it is possible to choose $\delta > 0$ such small that $|\omega'''(\theta)| \geq A$ for all $\theta \in \bigcup_{m=0}^{M} U_\delta(\theta_m)$ for some constant $A > 0$. Since $\omega''(\theta) = 0$ if and only if $\theta \in \Theta_{\text{cr}}$ there exists $B > 0$ with $|\omega''(\theta)| \geq B$ for all $\theta \in [0,\pi] \setminus \bigcup_{m=0}^{M} U_\delta(\theta_m)$. Now we write

$$g(t,c) = \int\limits_{\bigcup_{m=0}^{M} U_\delta(\theta_m)} \psi(\theta) \mathrm{e}^{\mathrm{i}t\phi(\theta,c)}\, \mathrm{d}\theta + \int\limits_{[0,\pi]\setminus\bigcup_{m=0}^{M} U_\delta(\theta_m)} \psi(\theta) \mathrm{e}^{\mathrm{i}t\phi(\theta,c)}\, \mathrm{d}\theta$$

and apply (3.23). Thus

$$|g(t,c)| \leq (M+1)\left(18A^{-1/3} + 8B^{-1/2}\right) C_\psi t^{-1/3}$$

holds for $t > 1$. Using $|g(t,c)| \leq \pi \max_{\theta\in[0,\pi]} |\psi(\theta)|$ in case $0 \leq t \leq 1$ proves the conclusion for $C_\omega = 2\max\{\pi, (M+1)(18A^{-1/3} + 8B^{-1/2})\}$. □

The next result provides the decay rate $t^{-1/2}$ along noncritical rays. The importance is to characterize the width of the regions around the critical rays with decay rate $t^{-1/3}$ that has to be excluded. This result provides sharp estimates for cross-over between the two decay rates. Excluding group velocities near the critical ones corresponding to the critical wave numbers, i.e. allowing only for c with $|c| \notin \bigcup_{\theta_n\in\Theta_{\text{cr}}} [|\omega'(\theta_n)| - \varepsilon, |\omega'(\theta_n)| + \varepsilon]$ where $\varepsilon > 0$, (3.25) implies a uniform bound $\sim t^{-1/2}$ on $g(t,c)$. In fact, the result shows that the excluded regions may be taken smaller, namely of width growing like $t^{2/3}$. Using a suitable Airy scaling, it can be shown that this width cannot be decreased, see (3.32) for more details.

Lemma 3.6 (Envelope function):
Consider the oscillatory integral (3.21) *with dispersion relation ω satisfying* (3.8) *and $\psi \in W^{1,1}([0,\pi])$. Then, there exists a constant $C_\omega > 0$ depending only on ω such that for all $t > 0$ and all $c \in \mathbb{R} \setminus \{c \mid \exists \theta \in \Theta_{\text{cr}} : ||\omega'(\theta)| - |c|| \leq t^{-2/3}\}$ holds*

$$|g(t,c)| \leq \frac{C_\omega C_\psi}{(1+t)^{1/2}} \left(1 + \sum_{\theta\in\Theta_{\text{cr}}} \frac{1}{|\omega'(\theta)^2 - c^2|^{1/4}}\right). \tag{3.25}$$

Proof. For $0 < t \leq 1$ we use $|g(t,c)| \leq C_\psi \pi$. Below we assume $t > 1$.

To simplify the considerations let us first assume that there is only one critical wave number $\theta_0 = 0$. For θ near 0 the phase function of (3.21) behaves like $\phi(\theta,c) = \pm(c_0 - c)\theta \pm \frac{\omega'''(0)}{6}\theta^3 + \mathcal{O}(\theta^5)$ with $c_0 = \omega'(0)$. Now we write

$$g(t,c) = \int_0^{\tilde\delta} \psi(\theta)\mathrm{e}^{\mathrm{i}t\phi(\theta,c)}\, \mathrm{d}\theta + \int_{\tilde\delta}^{\delta} \psi(\theta)\mathrm{e}^{\mathrm{i}t\phi(\theta,c)}\, \mathrm{d}\theta + \int_{\delta}^{\pi} \psi(\theta)\mathrm{e}^{\mathrm{i}t\phi(\theta,c)}\, \mathrm{d}\theta. \tag{3.26}$$

Due to $\omega'''(0) \neq 0$ there exists $0 < \delta < 1$ and constants $\underline{A}, \bar{A} > 0$ such that

$$\forall\, \theta \in (0,\delta): \qquad |\omega''(\theta)| \geq \underline{A}\theta \quad \text{and} \quad |\omega'(\theta) - c_0| \leq \bar{A}\theta^2. \tag{3.27}$$

Then we have in particular $|\partial_\theta^2 \phi(\theta, c)| = |\omega''(\theta)| \geq \underline{A}\tilde{\delta}$ for all $\theta \in (\tilde{\delta}, \delta)$. Since we assumed $\Theta_{\rm cr} = \{0\}$ there also exists $B > 0$ such that we have $|\partial_\theta^2 \phi(\theta, c)| \geq B$ for all $\theta \in (\delta, \pi)$. Thus van der Corput's Lemma (3.23) implies

$$\left|\int_{\tilde{\delta}}^{\delta} \psi(\theta) \mathrm{e}^{\mathrm{i}t\phi(\theta,c)}\, \mathrm{d}\theta\right| \leq \frac{8C_\psi}{(\underline{A}\tilde{\delta}t)^{1/2}} \quad \text{and} \quad \left|\int_{\delta}^{\pi} \psi(\theta) \mathrm{e}^{\mathrm{i}t\phi(\theta,c)}\, \mathrm{d}\theta\right| \leq \frac{8C_\psi}{(Bt)^{1/2}}. \tag{3.28}$$

Here $\delta, \underline{A}, \bar{A}$ and B do not depend on c but only on ω.

If $\tilde{\delta}$ with $0 < \tilde{\delta} \leq \delta$ is so small that $\tilde{\delta}^2 \leq \frac{1}{\bar{A}+1}|c_0 - |c||$, then using (3.27) we obtain $|\partial_\theta \phi(\theta, c)| = |\omega'(\theta) - c| \geq |c_0 - |c|| - \bar{A}\theta^2 \geq \tilde{\delta}^2$ for all $\theta \in (0, \tilde{\delta})$. Hence, again according to (3.23) we obtain

$$\left|\int_{0}^{\tilde{\delta}} \psi(\theta) \mathrm{e}^{\mathrm{i}t\phi(\theta,c)}\, \mathrm{d}\theta\right| \leq \frac{3C_\psi}{\tilde{\delta}^2 t}. \tag{3.29}$$

Now we distinguish two cases. If $\delta^2 \leq \frac{1}{\bar{A}+1}|c_0 - |c||$ we choose $\tilde{\delta} := \delta$. Hence the right hand side of (3.29) is independent of c. Substituting this bound together with the second estimate in (3.28) in (3.26) gives

$$|g(t,c)| \leq \frac{8C_\psi}{(Bt)^{1/2}} + \frac{3C_\psi}{\delta^2 t} \leq \frac{C_\psi}{t^{1/2}}\left(\frac{8}{\sqrt{B}} + \frac{3}{\delta^2}\right). \tag{3.30}$$

In case $\frac{1}{\bar{A}+1}|c_0 - |c|| < \delta^2$ we choose $\tilde{\delta}^2 := \frac{1}{\bar{A}+1}|c_0 - |c||$. Then the assumption $|c_0 - |c|| \geq t^{-2/3}$ yields $\tilde{\delta}^{3/2} t^{1/2} \geq (\bar{A}+1)^{-3/4}$. Thus, combining the upper bound (3.29) with the first estimate in (3.28) leads to

$$\left|\int_{0}^{\delta} \psi(\theta) \mathrm{e}^{\mathrm{i}t\phi(\theta,c)}\, \mathrm{d}\theta\right| \leq \frac{8C_\psi}{(\underline{A}\tilde{\delta}t)^{1/2}} + \frac{3C_\psi}{\tilde{\delta}^2 t} \leq \frac{C_\psi}{|c_0 - |c||^{1/4} t^{1/2}}\left(\frac{8}{\underline{A}^{1/2}} + 3(\bar{A}+1)^{3/4}\right).$$

Finally, using this together with the second estimate in (3.28) and $|g(t,c)| \leq C_\psi \pi$ for $0 < t \leq 1$ yields

$$|g(t,c)| \leq \frac{C_\omega C_\psi}{(1+t)^{1/2}}\left(1 + \frac{1}{|c_0^2 - c^2|^{1/4}}\right)$$

with $C_\omega > 0$ depending only on $\omega(\theta)$. The last estimate also covers (3.30) if we choose C_ω sufficiently large. This completes the proof for $\Theta_{\rm cr} = \{0\}$.

To prove the general case assume we have $\Theta_{\rm cr} = \{\theta_0, \theta_1, \dots, \theta_M\}$ with $\theta_0 < \theta_1 < \dots \theta_M$. We decompose the integral defining $g(c,t)$ like

$$g(c,t) = \dots + \int_{\theta_m}^{\theta_m + \tilde{\delta}_m} \dots + \int_{\theta_m + \tilde{\delta}_m}^{\theta_m + \delta_m} \dots + \int_{\theta_m + \delta_m}^{\theta_{m+1} + \delta_{m+1}} \dots + \dots$$

with δ_m and δ_{m+1} sufficiently small such that $\omega'''(\theta) \neq 0$ for $\theta \in (\theta_m - \delta_m, \theta_m + \delta_m) \cup (\theta_{m+1} - \delta_{m+1}, \theta_{m+1} + \delta_{m+1})$. Then similar estimates like (3.27) hold and we use the same arguments as above to get the upper bound

$$\Big| \int_{\theta_m}^{\theta_{m+1}-\delta_{m+1}} \psi(\theta) \mathrm{e}^{\mathrm{i}t\phi(\theta,c)} \,\mathrm{d}\theta \Big| \leq \frac{C_{\omega,m} C_\psi}{(1+t)^{1/2}} \left(1 + \frac{1}{|c_m^2 - c^2|^{1/4}}\right).$$

Since Θ_{cr} is finite this implies the statement. □

Now we state a result that provides a global decay rate $t^{-1/2}$ under the additional assumption that only $\theta = 0$ is a critical wave number and that the function ψ satisfies $\psi(0)$. It will be used to estimate $\mathrm{e}^{\mathcal{L}t} \mathcal{J}_{\mathrm{r}}$, where the bad behavior of the fronts, which relate to long waves (i.e. $\theta = 0$) are filtered out by the difference operators $\partial_\pm - 1$ in $\mathcal{J}_{\mathrm{r}}$.

Lemma 3.7:
Consider the oscillatory integral (3.21) *with dispersion relation* ω *satisfying* (3.8) *and* $\psi \in W^{1,1}([0,\pi])$. *If additionally* $\Theta_{\mathrm{cr}} = \{0\}$ *and* $\psi(0) = 0$, *then there exists a constant* $C_\omega > 0$ *depending only on* $\omega(\theta)$ *such that*

$$\forall t \geq 0: \quad \big|g(t,c)\big| \leq \frac{C_\omega C_\psi}{(1+t)^{1/2}}. \tag{3.31}$$

The proof relies on an uniform asymptotic expansion of the oscillatory integrals. Since we think that the technical details would dislocate the focus of the paper we forbear to give the full proof but only highlight the main idea. The detail can be found in [Pat09].

To see the filter effect of the difference operators $\partial_\pm - 1$ we apply the method of stationary phase, cf. [Won89], to $g(t,c)$ for $c = c_0 := \omega'(0)$ and find that it behaves like $t^{-2/3}$. According to [Hör90, 7.7.18] there is a generalization of the classical method of stationary phase which is uniform in terms of the group velocity c. In fact, for $y \in [-\varepsilon, \varepsilon]$ with $\varepsilon > 0$ sufficiently small and $c_0 := \omega'(0)$ holds

$$\begin{aligned} g(t, c_0 + y) \sim t^{-1/3} \mathrm{Ai}\big(a(y)t^{2/3}\big)\big[u_0(y) + \mathcal{O}(t^{-1})\big] \\ + t^{-2/3} \mathrm{Ai}'\big(a(y)t^{2/3}\big)\big[u_1(y) + \mathcal{O}(t^{-1})\big]. \end{aligned} \tag{3.32}$$

Here $\mathrm{Ai}(\cdot)$ refers to the Airy function, and a, u_0 and u_1 are smooth functions with $a(0) = 0$. Making these functions explicit we find that the leading order term cancels. Together with Lemma 3.6 this implies (3.31).

In this connection one should note that there is a smooth cross-over between the different scales. Indeed, employing the asymptotic behavior of Airy's function, cf. [Olv74], we obtain for $y < 0$ the asymptotic behavior $t^{-1/3}\mathrm{Ai}\big(a(y)t^{2/3}\big) \sim C_1 t^{-1/2}$ and $t^{-2/3}\mathrm{Ai}'\big(a(y)t^{2/3}\big) \sim C_2 t^{-1/2}$ as $t \to \infty$. Furthermore, the asymptotic expansion (3.32) implies that the width-scaling of the fronts in Lemma 3.5 is sharp. This holds for $\theta = 0$ as well as for general $\theta \in \Theta_{\mathrm{cr}}$, where in (3.32) occurs an additional modulating factor $\mathrm{e}^{i\omega(\theta)t}$, see again [Hör90] and [Pat09] for details.

Up to now we provided the decay rates along critical and noncritical rays but we did not use that the effective propagation speed is finite. The light cone corresponds to $c \in [-c_*, c_*]$ where $c_* := \max_{\theta\in\Theta_{\rm cr}} |\omega'(\theta)|$. Outside of this region the decay is faster than algebraic in terms of t as well as in terms of the velocity $c \in (-\infty, -c_*) \cup (c_*, \infty)$.

Applying partial integration, which is the standard argument to see this, cf. [Ste93], is not straight forward due to the occurring boundary terms. In [Fri03] the exponential decay is proved, for the standard FPU case, using a dilation-analytic argument with respect to Fourier frequency. Using (3.32) and the asymptotic behavior of Ai as $t \to \infty$ it turn out that the decay is even faster. In any case one finds for each $\delta > 0$ a decay constant $\kappa_\delta > 0$ such that

$$\forall t \geq 0 \; \forall c \in (-\infty, -c_*-\delta] \cup [c_*+\delta, \infty): \qquad |g(t,c)| \leq \mathrm{e}^{-\kappa_\delta(|c|-c_*)t}. \tag{3.33}$$

With the above lemmas we are now prepared to prove the ℓ^p decay rate of $\mathrm{e}^{\mathcal{L}t}$.

Proof of Theorem 3.1. According to Lemma 3.4 the group $\mathrm{e}^{\mathcal{L}t}$ acts as convolution with the matrix-valued Green's function $\mathbf{G}(t) = (\mathbf{G}^{k,m}(t))_{k,m=1,2}$. Using Young's inequality (3.6) we obtain

$$\|\mathrm{e}^{\mathcal{L}t}\mathbf{z}^0\|_{\ell^p} \leq \|\mathbf{G}(t)\|_{\ell^p}\|\mathbf{z}^0\|_{\ell^1}.$$

Thus, it is sufficient to prove the desired decay rates in (3.9) and (3.10), respectively, for the components of $\mathbf{G}(t)$.

We only carry out the details of the proof for $\mathbf{G}^{1,1}(t)$. Let us first consider the case $p \neq 4$. We aim to prove

$$\left\|\mathbf{G}^{1,1}(t)\right\|_{\ell^p} \leq \frac{C_p}{(1+t)^{\alpha_p}} \tag{3.34}$$

which according to (3.18) and by introducing the velocity $c = j/t$ as new variable follows from

$$t\int_{-\infty}^{\infty}\left|\frac{1}{2\pi}\int_0^{\pi} h(\theta,t,c)\,\mathrm{d}\theta\right|^p \mathrm{d}c = \mathcal{O}(t^{-p\alpha_p}) \quad \text{as } t \to \infty.$$

The left hand side is bounded by terms of the form

$$B_p(t) := t\int_{-\infty}^{\infty} |g(t,c)|^p \,\mathrm{d}c$$

with $g(t,c)$ defined in (3.21), $\phi(\theta,c) = \pm(\omega(\theta) \pm c\theta)$ and $\psi(\theta)$ standing for 1, $1/\omega_{\rm r}(\theta)$ or $\omega_{\rm r}(\theta)$. Without loss of generality we only consider $\phi(\theta,c) = \omega(\theta) - c\theta$ and may assume $t > 1$.

To estimate the contributions of each $\theta \in \Theta_{\mathrm{cr}}$ we choose $\varepsilon > 0$ and consider $c \in [\omega'(\theta)-\varepsilon, \omega'(\theta)+\varepsilon]$. Using Lemma 3.5 and Lemma 3.6 we find

$$\begin{aligned} B_p(t) &= t\left(\int_{\omega'(\theta)-\varepsilon}^{\omega'(\theta)-t^{-2/3}} + \int_{\omega'(\theta)-t^{-2/3}}^{\omega'(\theta)+t^{-2/3}} + \int_{\omega'(\theta)+t^{-2/3}}^{\omega'(\theta)+\varepsilon}\right) |g(t,c)|^p \,\mathrm{d}c \\ &\leq \frac{C_\omega C_\psi}{(1+t)^{p/3-1/3}} + \frac{2\tilde{C}_\omega C_\psi}{(1+t)^{p/2-1}}\left(1 + \int_{\omega'(\theta)+t^{-2/3}}^{\omega'(\theta)+\varepsilon} \frac{\mathrm{d}c}{|\omega'(\theta)^2 - c^2|^{p/4}}\right) \\ &\leq \frac{C_\omega C_\psi + 2\tilde{C}_\omega C_\psi C}{(1+t)^{(p-1)/3}} + \frac{2\tilde{C}_\omega C_\psi C}{(1+t)^{(p-2)/2}} \end{aligned} \tag{3.35}$$

with C depending on $\omega'(\theta), \varepsilon$ and p. Taking the leading order term we get the decay rate $p\alpha_p$. Thus, using (3.33) and Lemma 3.6 for $c \notin [\omega'(\theta)-\varepsilon, \omega'(\theta)+\varepsilon]$ we obtain

$$B_p(t) \leq 2M\frac{C_\omega C_\psi + 2\tilde{C}_\omega C_\psi C}{(1+t)^{p\alpha_p}} + \mathcal{O}(t^{-(p-2)/2}) + \mathcal{O}(\mathrm{e}^{-\kappa_\varepsilon \varepsilon p t})$$

which implies (3.34). Hence, the case $p \neq 4$ is established.

In the case $p = 4$ the additional factor $\log t$ contributing to the leading order term appears on the right hand side of (3.35). Indeed, we obtain

$$B_4(t) \leq \frac{C_\omega C_\psi + 2\tilde{C}_\omega C_\psi C}{1+t} + \frac{2\tilde{C}_\omega C_\psi C}{1+t}\Big(\log t + \log \varepsilon\Big).$$

This is sufficient to see that $\left\|\mathbf{G}^{1,1}(t)\right\|_{\ell^p} \leq C_p\big((1+t)\log(2+t)\big)^{1/4}$.

For the other components of $\mathbf{G}(t)$ we may use exactly the same arguments. This proves the first statement of Theorem 3.1.

To prove the second statement we proceed like above but we use the global upper bound Lemma 3.7 instead of Lemma 3.5 and Lemma 3.6. Then, the leading order term behaves like $t^{(2-p)/p}$. □

4 Outlook: Further applications

4.1 Discrete Klein-Gordon and nonlinear Schrödinger equations

Here we outline how to apply the tools developed in Sections 2 and 3 to other models in one-dimensional chains, namely the discrete Klein-Gordon (dKG) and the discrete nonlinear Schrödinger equation (dNLS), see Section 1.

For (dKG) we have an on-site potential with $W'(x) = bx + \mathcal{O}(|x|^\beta)$. Like in the FPU case our results are not restricted to nearest neighbor interaction. Indeed, we may allow for any finite-range interaction as long as the stability condition (3.3) and the non-dependeracy condition (3.8) are satisfied; but for simplicity we restrict ourselves to the simplest case, where the dispersion relation reads

$$\omega(\theta) = \sqrt{2 + b - 2\cos\theta}.$$

The stability condition immediately implies $b \geq 0$. In Figure 4.1 we plot the dispersion relation and the time evolution of a prototypical dKG chain. A major

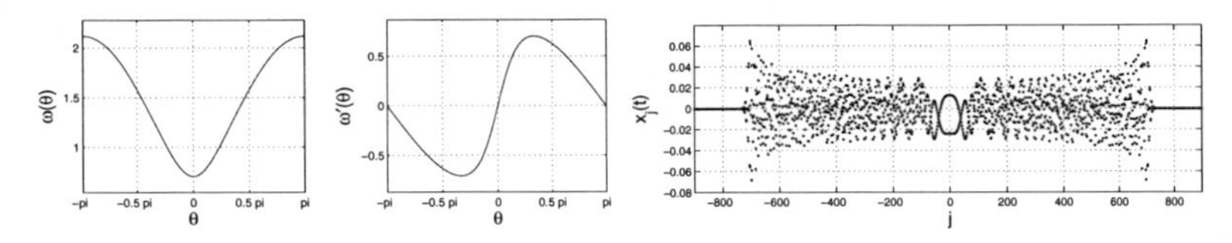

Figure 4.1: Dispersion relation and time evolution for the prototypical dKG chain ($a_1 = -1$, $b = 0.5$): $\omega(\theta)$, $\omega'(\theta)$ and $x_j(t)$ at $t = 800$ to initial condition $(x_j(0), \dot{x}_j(0)) = (\delta_{j,0}, 0)$.

difference to FPU is that the propagation fronts do not correspond to the macroscopic wave number $\theta \approx 0$. Hence, the fronts are not monotone but have an Airy expansion as in (3.32) but multiplied with a factor $\mathrm{e}^{\mathrm{i}\omega(\theta_*)t}$, where $\omega'(\theta_*) = c_*$ and hence $\omega''(\theta_*) = 0$. Now $\theta = 0$ does not lie in Θ_{cr} because the on-site potential W destroyed the Galilean invariance.

But apart from these two difference the results and the approaches to prove these are the same like in the FPU case. Using the explicit solution of the linearized system, we may prove the analog to Theorem 3.1 with the same decay rates. This relies on the fact that the key ingredients for its proof is the representation of the solutions in terms of oscillatory integral of the form (3.21) and quite general conditions on ω, namely (3.8).

Theorem 4.1:
Consider the discrete Klein-Gordon system (dKG) *with* $W'(x) = bx + \mathcal{O}(|x|^\beta)$, $b < 0$ *and* $\beta > 4$. *Then, for each* $p \in [2, 4) \cup (4, \infty]$ *there exist* C_p *and* $\varepsilon > 0$ *such that all solutions* $\mathbf{z} = (\mathbf{x}, \dot{\mathbf{x}})$ *with* $\|\mathbf{z}(0)\|_{\ell^1} \leq \varepsilon$ *satisfy the estimate*

$$\|\mathbf{z}(t)\|_{\ell^p} \leq \frac{C_p}{(1+t)^{\alpha_p}} \|\mathbf{z}(0)\|_{\ell^1} \quad \text{for all } t \geq 0, \tag{4.1}$$

where the decay rate α_p *is given in* (3.9).

Again the case $p = 4$ can be included by adding a suitable logarithmic term.

This theorem improves the result in [SK05] in a twofold manner, namely in terms of β as well as in terms of the decay rate α_p for $p \in (2, \infty)$. In particular, Theorem 4.1 explains the discrepancy between the numerical simulation and the theoretical decay rate $\hat{\alpha}_p$ in [SK05]. We see that our decay rate α_p fits the numerics much better.

p	4	5	6
$\hat{\alpha}_p = \frac{p-2}{3p}$	$\frac{1}{6} \approx 0.167$	$\frac{1}{5} = 0.2$	$\frac{2}{9} \approx 0.222$
numerics in [SK05]	0.226	0.267	0.292
$\alpha_p = \frac{p-1}{3p}$	$\frac{1}{4} = 0.25$ log	$\frac{4}{15} \approx 0.267$	$\frac{5}{18} \approx 0.278$

We recall that for the case $p = 4$ the optimal decay is like $(1+t)^{-1/4}\log(2+t)$. Hence, it is not surprising that the numerical approximation in this case is especially bad.

The above theory can be easily transferred to the discrete nonlinear Schrödinger equation (dNLS), where the dispersion relation reads $\omega(\theta) = 2-2\cos\theta$. Obviously $\Theta_{\mathrm{cr}} = \{\pi/2\}$ and the non-degeneracy condition $\omega'''(\pi) \neq 0$ holds. In this case the ℓ^2 norm is in fact a first integral, and hence is preserved exactly along solutions. Using this, it is not difficult to show that for $\beta > 4$ we have dispersive stability with the same decay rates as above.

4.2 Applications to systems in 2D

Here we discuss the application of our general theory to a system on a two-dimensional lattice. The crucial point in higher space dimensions are the estimates for the linear group. Here we only present a conjecture for the decay rates; the rigorous proof being ongoing work, cf. [Pat09]. For methods to handle 2D oscillatory integrals we refer to [Won89, BH86, Hör90] and [GWF81], which is based on techniques derived in [Dui74].

We consider the Hamiltonian system

$$\ddot{x}_j = V'(x_{j+e_1} - x_j) - V'(x_j - x_{j-e_1}) + V'(x_{j+e_2} - x_j) - V'(x_j - x_{j-e_2}) \tag{4.2}$$

with $j = (j_1, j_2) \in \mathbb{Z}^2$. Here $e_1 = (1,0)^T$ and $e_2 = (0,1)^T$ are the unit vectors, $\mathbf{x} := (x_j)_{j\in\mathbb{Z}^2}$ with $x_j \in \mathbb{R}$ and $V'(r) = r + \mathcal{O}(|r|^\beta)$ with $\beta > 1$. To avoid difficulties by introducing an analog to the distances $\mathbf{r}$ in one dimension we restrict ourselves to initial conditions $(\mathbf{x}(0), \dot{\mathbf{x}}(0)) = (\mathbf{x}^0, \mathbf{0}) \in \ell^1(\mathbb{Z}^2, \mathbb{R}^2)$.

Like in the one-dimensional case it is possible to solve the linearization of (4.2) explicitly and the behavior of the solutions relies on oscillatory integrals of the form

$$g(t,c) = \frac{1}{(2\pi)^2}\int_{\mathbb{T}^2} \psi(\theta)\mathrm{e}^{\mathrm{i}t\phi(\theta,c)}\,\mathrm{d}\theta \tag{4.3}$$

with $\phi(\theta, c) = \pm(\omega(\theta) - c\cdot\theta)$, where now $\theta = (\theta_1, \theta_2) \in \mathbb{T}^2$ and $c \in \mathbb{R}^2$. For (4.2) the dispersion relation is given by

$$\omega(\theta) = \sqrt{4 - 2\cos\theta_1 - 2\cos\theta_2}.$$

Although we do not state the formula note that in this case it is possible calculate the critical set $\Theta_{\mathrm{cr}} = \{\theta \in \mathbb{T}^2 \,|\, \det \mathrm{D}^2\omega(\theta) = 0\}$ explicitly. The mapping $\mathrm{D}\omega : \mathbb{T}^2 \to \mathbb{R}^2$ has the range $\{c \in \mathbb{R}^2 \,|\, 0 < |c| < 1\}$ of possible group velocities and maps Θ_{cr} into a closed curve with four vertices, see Figure 4.2, left. The right-hand side of Figure 4.2 displays the time evolution of the first component of the Green's function, which clearly shows different regimes at the critical wave numbers. We can roughly distinguish three regions: (i) four vertices, (ii) four edges connecting these vertices and (iii) the remaining region inside the light cone, which is a circle of radius t.

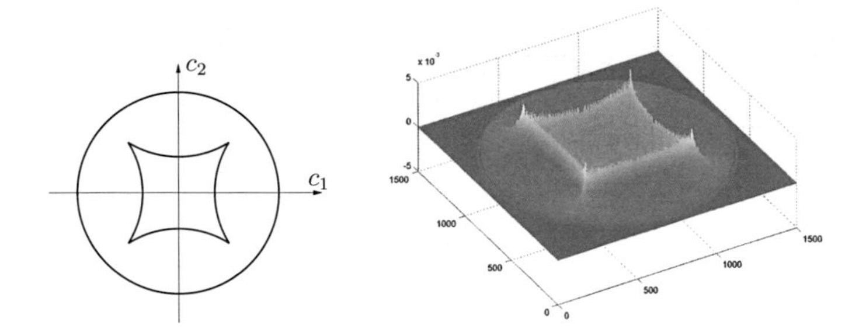

Figure 4.2: Left: The circle is the set of possible group velocities and the curve with four vertices denotes the critical group velocities. Right: Time evolution of the linearization of (4.2) with initial condition $x^0_{j_1 j_j} = \delta_{j_1}\delta_{j_2}$ and $\dot{\mathbf{x}}^0 = \mathbf{0}$.

To obtain the decay properties of $\|e^{\mathcal{L}t}\|_{\ell^1,\ell^p}$, where $\mathcal{L}$ again stands for the linear part of the operator on the right hand side of (4.2), we first determine the local asymptotic behavior of (4.3). Then, assuming a reasonable width of the three different regions we infer the ℓ^p decay rate like in the proof of Theorem 3.1. To do so we apply the localization principle: For $c = \mathrm{D}\omega(\theta)$, the main contribution to $g(t,c)$ is given by

$$I_\theta(t) = \int_{\mathbb{T}^2} h(\delta)\, e^{it\varphi(\delta)}\, d\delta \quad \text{where } \varphi(\delta) := \phi(\theta+\delta, \mathrm{D}\omega(\theta)). \tag{4.4}$$

Here $h \in C^\infty$, $\operatorname{supp} h \subset U_\varepsilon(0)$ for $\varepsilon > 0$ sufficiently small, and $h(0) = 1$. The function h arises via partition of unity on $\mathbb{T}^2$. The local decay rate $t^{-\alpha(\theta)}$ of $I_\theta(t)$, and hence of $g(t, \mathrm{D}\omega(\theta))$, is determined by the leading-order terms of the Taylor expansion

$$\varphi(\delta) = \varphi(0) + \frac{1}{2}\delta^T \cdot \mathrm{D}^2\omega(\theta)\cdot\delta + \text{h.o.t.}, \quad \mathrm{D}\varphi(0) = 0. \tag{4.5}$$

For $\theta \in \mathbb{T}^2 \setminus \Theta_{\text{cr}}$ a linear coordinate transformation $\delta = A\xi$ leads to $\varphi(A\xi) = \varphi(0) + \xi_1^2 \pm \xi_2^2 + \text{h.o.t.}$ Thus, scaling ξ with $\sqrt{t}$ leads to $|I_\theta(t)| \sim \frac{2\pi}{|\det \mathrm{D}^2\omega(\theta)|^{1/2}} t^{-1} + \mathcal{O}(t^{-2})$. This decay rate corresponds to the region inside the light cone, but away from the fronts.

For $\theta \in \Theta_{\text{cr}}$ we have to distinguish two cases. The four vertices correspond to the degenerated points $(\pm\frac{\pi}{2}, \pm\frac{\pi}{2})$. If $\theta \in \Theta_{\text{cr}} \setminus \{(\pm\frac{\pi}{2}, \pm\frac{\pi}{2})\}$, then, following the ideas in [GWF81] we find a local coordinate transformation to get $\varphi(A\xi) = \varphi(0) + \xi_1^2 \pm \xi_2^3 + \text{h.o.t.}$ Thus, $|I_\theta(t)| \sim b(\theta)t^{-5/6} + \mathcal{O}(t^{-7/6})$, where $b(\theta)$ is singular in $(\pm\frac{\pi}{2}, \pm\frac{\pi}{2})$.

Finally, for $\theta = (\pm\frac{\pi}{2}, \pm\frac{\pi}{2})$ there exists a coordinate transform $\delta = b(\xi)$ such that $\phi(b(\xi)) = \phi(0) - \xi_1^2 - \xi_2^4$. Scaling ξ_1 and ξ_2 with $t^{1/2}$ and $t^{1/4}$, respectively, gives $|I_\theta(t)| \sim b(\theta)t^{-3/4}$, which is also the global ℓ^∞ decay rate.

The decay rate of $\|\mathrm{e}^{\mathcal{L}t}\|_{\ell^1,\ell^p}$ is roughly determined by

$$\left\|g\left(t, \tfrac{\cdot}{t}\right)\right\|^p_{\ell^p(\mathbb{Z}^2,\mathbb{R})} \sim t^2 \int_{|c|\leq 1} |g(t,c)|^p \mathrm{d}c = \frac{t^2}{(2\pi)^{2p}} \int_{|c|<1} \left|\int_{\mathbb{T}^2} \mathrm{e}^{\mathrm{i}t\phi(\theta,c)}\,\mathrm{d}\theta\right|^p \mathrm{d}c. \quad (4.6)$$

Using the normal forms given above we estimate the amount of the three regions on the right-hand side and obtain

$$\left\|g(t, \tfrac{\cdot}{t})\right\|^p_{\ell^p} \sim t^2\Big(C_{\text{cone}}t^{-p} + C_{\text{curve}}t^{-\beta}t^{-5p/6} + C_{\text{vertex}}t^{-\gamma}t^{-3p/4}\Big),$$

where $t^{-\beta}$ gives area of the regions around the four curves and $t^{-\gamma}$ the area of the regions around the four vertices measured relatively to the disc $|c| < 1$. We conjecture that the correct values are $\beta = 2/3$ and $\gamma = 3/4$.

This conjecture leads to the decay rates $\alpha_p^{2\mathrm{D}} = \min\{\frac{p-2}{p}, \frac{3p-5}{4p}\}$, which is obtained from interpolating the three values $\alpha_2 = 0$, $\alpha_3 = 1/3$, and $\alpha_\infty = 3/4$. It seems reasonably that the case $p = 3$ needs a logarithmic correction. The numerical simulations shown in Figure 4.3 agree quite well with this rate for $p \in [2,3]$, however there are major discrepancies for larger p. In any case, the numerics clearly suggests that the optimal decay rates are better that the ones, which can be obtained by interpolating between $\alpha_2 = 0$ and $\alpha_\infty = 3/4$.

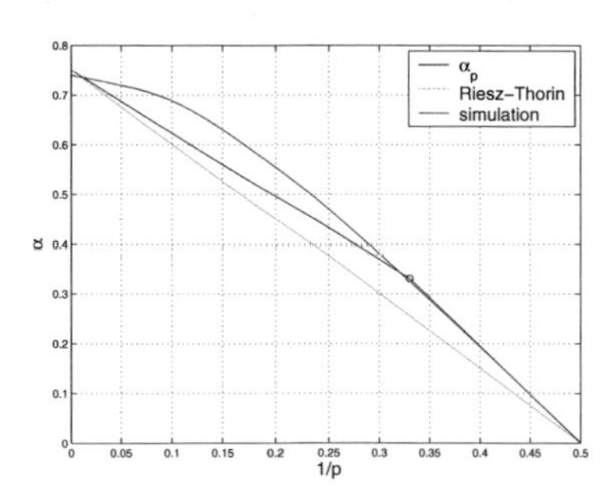

Figure 4.3: Conjectured exact decay rate $\alpha 2\mathrm{D}_p$, interpolation rate and numerically estimates rates as function of $1/p$.

Nevertheless, if $\alpha_p^{2\mathrm{D}}$ hat the above form for $p \in [2,3)$, then the nonlinear dispersive decay theory of Theorem 2.1 provides dispersive decay with this rate whenever the nonlinearity is of degree $\beta > 3$.

References

[BH86] N. Bleistein and R. A. Handelsman. *Asymptotic expansions of integrals.* Dover Publications Inc., New York, second edition, 1986.

[CW91] F. M. Christ and M. I. Weinstein. Dispersion of small amplitude solutions of the generalized Korteweg-de Vries equation. *J. Funct. Anal.*, 100(1), 87–109, 1991.

[Dui74] J. J. Duistermaat. Oscillatory integrals, Lagrange immersions and unfolding of singularities. *Comm. Pure Appl. Math.*, 27, 207–281, 1974.

[FP99] G. Friesecke and R. L. Pego. Solitary waves on FPU lattices. I. Qualitative properties, renormalization and continuum limit. *Nonlinearity*, 12(6), 1601–1627, 1999.

[Fri03] G. Friesecke. Dynamics of the infinite harmonic chain: conversion of coherent initial data into synchronized binary oscillations. *Preprint*, 2003.

[FW94] G. Friesecke and J. A. D. Wattis. Existence theorem for solitary waves on lattices. *Comm. Math. Phys.*, 161(2), 391–418, 1994.

[GHM06] J. Giannoulis, M. Herrmann, and A. Mielke. Continuum descriptions for the dynamics in discrete lattices: derivation and justification. In *Analysis, modeling and simulation of multiscale problems*, pages 435–466. Springer, Berlin, 2006.

[GWF81] A. D. Gorman, R. Wells, and G. N. Fleming. Wave propagation and thom's theorem. *J. Phys. A: Math. Gen.*, 14(7), 1519–1531, 1981.

[Hör90] L. Hörmander. *The analysis of linear partial differential operators. I*, volume 256 of *Grundlehren der Mathematischen Wissenschaften.* Springer-Verlag, Berlin, second edition, 1990.

[IJ05] G. Iooss and G. James. Localized waves in nonlinear oscillator chains. *Chaos*, 15(1), 015113–+, March 2005.

[IZ09] L. I. Ignat and E. Zuazua. Numerical dispersive schemes for the nonlinear Schrödinger equation. *SIAM J. Numer. Anal.*, 47(2), 1366–1390, 2009.

[Mie06] A. Mielke. Macroscopic behavior of microscopic oscillations in harmonic lattices via Wigner-Husimi transforms. *Arch. Ration. Mech. Anal.*, 181(3), 401–448, 2006.

[MSU01] A. Mielke, G. Schneider, and H. Uecker. Stability and diffusive dynamics on extended domains. In B. Fiedler, editor, *Ergodic Theory, Analysis, and Efficient Simulation of Dynamical Systems*, pages 563–583. Springer–Verlag, 2001.

[Olv74] F. W. J. Olver. *Asymptotics and special functions.* Academic Press, New York-London, 1974. Computer Science and Applied Mathematics.

[Pat09] C. Patz. *Dynamics of Hamiltonian Systems on Infinite Lattices.* PhD thesis, Humboldt University of Berlin, 2009. In preparation.

[Ree76] M. Reed. *Abstract non-linear wave equations.* Lecture Notes in Mathematics, Vol. 507. Springer-Verlag, Berlin, 1976.

[Seg68] I. Segal. Dispersion for non-linear relativistic equations. II. *Ann. Sci. École Norm. Sup. (4)*, 1, 459–497, 1968.

[SK05] A. Stefanov and P. G. Kevrekidis. Asymptotic behaviour of small solutions for the discrete nonlinear Schrödinger and Klein-Gordon equations. *Nonlinearity*, 18(4), 1841–1857, 2005.

[Ste93] E. M. Stein. *Harmonic analysis: real-variable methods, orthogonality, and oscillatory integrals*, volume 43 of *Princeton Mathematical Series.* Princeton University Press, Princeton, NJ, 1993.

[Str74] W. A. Strauss. Dispersion of low-energy waves for two conservative equations. *Arch. Rational Mech. Anal.*, 55, 86–92, 1974.

[Str78] W. A. Strauss. Nonlinear invariant wave equations. In *Invariant wave equations (Proc. "Ettore Majorana" Internat. School of Math. Phys., Erice, 1977)*, volume 73 of *Lecture Notes in Phys.*, pages 197–249. Springer, Berlin, 1978.

[SW00] G. Schneider and C. E. Wayne. Counter-propagating waves on fluid surfaces and the continuum limit of the Fermi-Pasta-Ulam model. In B. Fiedler, K. Gröger, and J. Sprekels, editors, *International Conference on Differential Equations*, volume 1, pages 390–404. World Scientific, 2000.

[Won89] R. Wong. *Asymptotic approximations of integrals.* Computer Science and Scientific Computing. Academic Press Inc., Boston, MA, 1989.

[Zua05] E. Zuazua. Propagation, observation, control and numerical approximation of waves approximated by finite difference method. *SIAM Review*, 47, 197–243, 2005.

Chapter 3

Uniform asymptotic expansions for the infinite harmonic chain

The following paper by A. MIELKE and C. PATZ is available as *WIAS Preprint No. 1846, Berlin 2013.*

We study the dispersive behavior of waves in linear oscillator chains. We show that for general general dispersions it is possible to construct an expansion such that the remainder can be estimated by $1/t$ uniformly in space. In particular we give precise asymptotics for the transition from the $1/t^{1/2}$ decay of non-degenerate wave numbers to the generate $1/t^{1/3}$ decay of generate wave numbers. This involves a careful description of the oscillatory integral involving the Airy function.

Uniform Asymptotic Expansions for the Infinite Harmonic Chain*

Alexander Mielke[1,2], Carsten Patz[2]

[1] Weierstraß-Institut für Angewandte Analysis und Stochastik
Mohrenstraße 39, 10117 Berlin, Germany

[2] Humboldt-Universität zu Berlin, Institut für Mathematik
Rudower Chaussee 25, 12489 Berlin-Adlershof, Germany

26 September 2013

Abstract

We study the dispersive behavior of waves in linear oscillator chains. We show that for general general dispersions it is possible to construct an expansion such that the remainder can be estimated by $1/t$ uniformly in space. In particular we give precise asymptotics for the transition from the $1/t^{1/2}$ decay of non-degenerate wave numbers to the generate $1/t^{1/3}$ decay of generate wave numbers. This involves a careful description of the oscillatory integral involving the Airy function.

*This research was partially supported by ERC through ERC-AdG 267802 AnaMultiScale

Contents

1 Introduction

In this work we study the dispersive behavior of waves in linear oscillator chains. While there is a large body of the analysis in certain regimes of the dispersion relation there seems to be no general theory providing a uniform estimates. The main problem derives from the fact that the large-time asymptotics of the solutions can be estimated along the rays given by the group velocity by the presentation via oscillatory integrals. The difficulty to obtain uniform estimates for the remainders of an asymptotic expansion stems from the fact that the dispersion relation $\theta \mapsto \omega(\theta)$ necessarily contains degenerate points (i.e. where $\omega''(\hat{\theta}) = 0$). While for non-degenerate points the solutions decay like $t^{-1/2}$, the decay at degenerate points the decay is only of order $t^{-1/k}$ with $k \geq 3$. Such separate estimates for dispersive partial differential equations or discrete lattices are classical (cf. [Whi74, Hör90, Ste93, FP99, Fri03, IZ05, SK05, MP10] and the references there), but our aim is to find a uniform expansion providing also a sharp estimate in the transition regions, i.e. for nearly degenerate wave numbers. Moreover, for non-degenerate wave numbers the asymptotic profiles are given in terms of simple trigonometric functions, the degenerate case with $k = 3$ (i.e. $\omega'''(\hat{\theta}) \neq 0$) leads to fronts with a profile given in terms of the Airy function, see Figure 1.1.

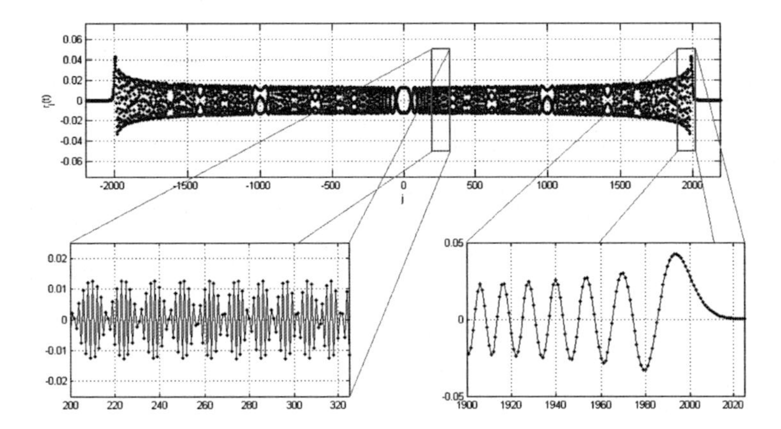

Figure 1.1: Green's function $G_{11}(t, j)$ at $t = 2000$ for the linear FPU chain. Lower left: periodic wave trains for non-degenerate group velocities. Lower right: Airy-type behavior at the degenerate front.

To be more precise we study the general linear oscillator chain

$$\ddot{x}_j = -a_0 x_j + \sum_{k=1}^{K} a_k \big(x_{j+k} - 2x_j + x_{j-k} \big), \quad j \in \mathbb{Z}, \tag{1.1}$$

where $a_0 \geq 0$ is due a stabilizing background potential and $a_1, ..., a_K \in \mathbb{R}$ give the interaction coefficients. With $a_0 = 0$, $a_1 = 1$, and $K = 1$ we obtain the linearized Fermi-Pasta-Ulam (FPU) system (cf. [FPU55]). The dispersion relation is given via

$$\omega^2 = \Lambda(\theta) := a_0 + \sum_{k=1}^{K} 2a_k \left[1 - \cos(k\theta)\right] \quad \text{for } \theta \in \mathcal{S}^1 := \mathbb{R}/_{(2\pi\mathbb{Z})},$$

where we always assume stability in the form $\Lambda(\theta) \geq 0$. Thus, we define the positive branch of the dispersion relation and the group velocity via

$$\omega(\theta) := \sqrt{\Lambda(\theta)} \geq 0 \quad \text{and} \quad c(\theta) := \omega'(\theta).$$

The oscillatory integrals to be estimated have the form

$$\int_{\mathcal{S}^1} A(\theta) \mathrm{e}^{\mathrm{i}(\omega(\theta)t + \theta j)} \, \mathrm{d}\theta = \int_{\mathcal{S}^1} A(\theta) \mathrm{e}^{\mathrm{i}(\omega(\theta) + c\theta)t} \, \mathrm{d}\theta \;=:\; g_A(t,c),$$

where we always use the relation $j = ct \in \mathbb{Z}$, which is important to keep the periodicity of the integrand.

Two of the analytical difficulties in the analysis can be explained at this point. First, if $\Lambda(\theta) = \gamma^2(\theta - \theta_1)^{2n} + O(|\theta - \theta_1|^{2n+1})$ the dispersion relation $\omega(\theta)$ may be non-analytic. We treat this case by assuming $a_0 = 0$, which gives $\omega(\theta) = \gamma|\theta| + O(\theta^2)$ where we assume $\gamma = \left(\sum_{k=1}^{K} k^2 a_k\right)^{1/2} > 0$. The second difficulty arises from degeneracies of the dispersion relation. The group velocity is given by the relation

$$c = c_{\mathrm{gr}}(\theta) = \omega'(\theta).$$

Fixing a θ with $\omega''(\theta) \neq 0$ and thus fixing $c = \omega'(\theta)$ and assuming that the support of the A in the definition of g_A is contained in a sufficiently small neighborhood of θ, we have the expansion

$$g_A(t,c) = c_A \cos\left[(\omega(\theta) + c\theta)t + \mathrm{sign}(\omega''(\theta))\frac{\pi}{4}\right] t^{1/2} + R_A^{\mathrm{non}}(t,c)$$

with $R_A^{\mathrm{non}}(t,c) = O(t^{-1})$, see Section 4.2. However, for θ near a degenerate case with $\omega''(\hat{\theta}) = 0$ but $\omega'''(\theta) \neq 0$, we obtain the expansion

$$g_A(t,\hat{c}) = \mathrm{e}^{\mathrm{i}tb(c)} \left[c_A \mathrm{Ai}(a(c)t^{2/3})t^{-1/3} + d_A \mathrm{Ai}'(a(c)t^{2/3})t^{-2/3}\right] + R_A^{\mathrm{deg}}(t,c)$$

with $R_A^{\mathrm{deg}}(t,c) = O(t^{-1})$ and suitable functions $a(c)$ and $b(c)$, see [Hör90] or Section 4.3. To obtain a uniform error estimate we give quantitative estimates for the two error terms $R_A^{\mathrm{non}}(t,c)$ and $R_A^{\mathrm{deg}}(t,c)$ and show that the degenerate expansion based on the Airy function coincides up to an error $O(t^{-1})$ with the harmonic expansion in an overlapping region.

In summary our main result reads as follows.

Theorem 1.1:
Assume that (1.1) *satisfies* $a_0 = 0$ *and the dispersion relation has the form* $\omega(\theta) = \operatorname{sign}(\theta)\tilde{\omega}(\theta)$, *where* $\tilde{\omega}$ *is smooth and satisfies the*

$$\text{non-degeneracy condition: } \tilde{\omega}''(\theta) = 0 \quad \Longrightarrow \quad \tilde{\omega}'''(\theta) \neq 0.$$

Then there exists a constant $C(\tilde{\omega})$ *such that for all* $t > 0$ *the Green's function matrix* $G_j(t) \in \mathbb{R}^{2\times 2}$ *for* (1.1) *written for the vectors* $\mathbf{r} = (x_{j+1}-x_j)_{j\in\mathbb{Z}}$ *and* $\mathbf{p} = (\dot{x}_j)_{j\in\mathbb{Z}}$ *satisfies the estimate*

$$|G_j(t) - \mathcal{G}^{\text{expan}}(t, j/t)| \leq C(\tilde{\omega})/t \quad \text{for all } j \in \mathbb{Z} \text{ and all } t > 0,$$

where the function $\mathcal{G}^{\text{expan}}(t, c)$ *is given in* (3.8).

Using the classical decay estimates for the Green's functions for group velocities outside the range of $\tilde{\omega}'$ (see Proposition 2.3), we easily obtain bounds in ℓ^p spaces, namely for each $p > 1$ there exists $C_p > 0$ such that

$$\|G_{(\cdot)}(t) - \mathcal{G}^{\text{expan}}(t, \cdot/t)\|_{\ell^p} \leq C_p t^{-(p-1)/p} \quad \text{for } t > 0.$$

In particular, this result implies that the dispersive decay estimates for the Green's function given in [MP10] for $p \in (2, 4) \cup (4, \infty)$

$$\|G_{(\cdot)}(t)\|_{\ell^p} \leq C_p^{\text{upper}} t^{-\alpha_p} \text{ for } t > 0 \quad \text{with } \alpha_p = \min\{\tfrac{p-2}{2p}, \tfrac{p-1}{3p}\}$$

are sharp. Our hope is that using the specific form of $G_j^{\text{expan}}(t)$ one can improve the decay results for nonlinear systems as well, see [GHM06, SK05, MP10].

Our analysis was stimulated by the work in [Fri03], which analyzed synchronization effects occurring for $c = 0$, which corresponds to the wave number $\theta = \pm\pi$. The analysis is done for fixed j and can be made uniform on parabolic regions $j^2 \leq Ct$. The dispersion of the energy was analyzed in [Mie06, HLTT08] via the Husimi and Wigner transform, even in multidimensional cases. The usage of dispersion in the error control for discretized PDEs is discussed in [IZ05, Ign07].

Notations. Bold face letters $\mathbf{r}, \mathbf{p}, \ldots$ denote elements in $\ell^p(\mathbb{R})$ or, with the common abuse of notation, smooth functions $\mathbf{r}(\cdot), \mathbf{p}(\cdot), \cdots : \mathbb{R} \to \ell^p(\mathbb{R})$.

Capital letters normally refer to linear operators, e.g. $G_j(t) : \mathbb{R}^2 \to \mathbb{R}^2$ and $\mathbf{G}(t) : \ell^p(\mathbb{R}^2) \to \ell^p(\mathbb{R}^2)$ or, again, to smooth functions mapping into these spaces.

To simplify the notations we denote bounds in general by C and carry out the distinction via indices $C_1, C_2, \ldots$ only if it is necessary. To highlight the dependency on parameters we write for instance $C = C(\omega, \delta)$. For $\delta \in \mathbb{R}$ the notation is obvious. For ω being a sufficiently smooth function this refers to a dependency on $\|\omega\|_{W^{n,p}}$ for suitable $n \in \mathbb{N}_0$ and $p \in \mathbb{N}$.

2 Dispersion in the generalized linear FPU

2.1 The generalized linear FPU

We consider an infinite number of equal particles with unit mass interacting with a finite number K of neighbors via linear forces. According to Newton's law, the equations of motion are

$$\ddot{x}_j = \sum_{1\le k\le K} \big[a_k(x_{j+k} - x_j) - a_k(x_j - x_{j-k})\big], \qquad j \in \mathbb{Z}. \tag{2.1}$$

Here $x_j \in \mathbb{R}$ denote the displacements. We write $\mathbf{x} := (x_j)_{j\in\mathbb{Z}}$. The system (2.1) is Hamiltonian, i.e. $(\dot{\mathbf{x}}, \dot{\mathbf{p}})^T = \mathcal{J}_{\mathsf{can}}\, \mathrm{d}\,\mathcal{H}_{\mathrm{x}}(\mathbf{x}, \mathbf{p})^T$ with momentum $\mathbf{p} := \dot{\mathbf{x}}$, Hamiltonian function $\mathcal{H}_{\mathrm{x}}(\mathbf{x}, \mathbf{p}) = \sum_{j\in\mathbb{Z}} \big(\frac{1}{2}p_j^2 + \sum_{1\le k\le K} \frac{a_k}{2}(x_{j+k} - x_j)^2\big)$ and $\mathcal{J}_{\mathsf{can}}$ the canonical Poisson operator defined by $\langle(\mathbf{x}, \mathbf{p})^T, \mathcal{J}_{\mathsf{can}}(\tilde{\mathbf{x}}, \tilde{\mathbf{p}})^T\rangle_{\ell^2\oplus\ell^2} = \langle\mathbf{x}, \tilde{\mathbf{p}}\rangle_{\ell^2} - \langle\tilde{\mathbf{x}}, \mathbf{p}\rangle_{\ell^2}$.

The system (2.1) exhibits plane waves solution of the form $x_j(t) = \mathrm{e}^{\mathrm{i}(\theta j + \hat{\omega} t)}$ if and only if the dispersion relation

$$\hat{\omega}^2 = \Lambda(\theta) := \sum_{1\le k\le K} 2a_k\big[1 - \cos(k\theta)\big] \tag{2.2}$$

is satisfied. By periodicity, it suffices to take $\theta \in (-\pi, \pi]$. We have $\Lambda(0) = 0$ which is a consequence of Galilean invariance, i.e. for all $\xi, c \in \mathbb{R}$ the transformation $(x_j, p_j) \mapsto (x_j{+}\xi{+}ct, p_j{+}c)$ leaves (2.1) invariant. Throughout, we assume the stability condition

$$\forall\, \theta \in (-\pi, \pi] \setminus \{0\} : \quad \Lambda(\theta) > 0 \tag{2.3}$$

holds. This certainly holds if all a_k are positive, however more general cases are possible. Thus we are able to define the relevant branch $\hat{\omega} = \omega(\theta)$ of the dispersion relation via

$$\omega(\theta) := \sqrt{\Lambda(\theta)} \ge 0. \tag{2.4}$$

With a slight abuse of notation we simply call ω the dispersion relation.

Due to the Galilean invariance it is convenient to use distances instead of the displacements $\mathbf{r} := (\partial_1 - \mathbf{1})\mathbf{x} = (x_{j+1} - x_j)_{j\in\mathbb{Z}}$ as new variables. Then the Hamiltonian function turns into $\mathcal{H}_{\mathrm{r}}(\mathbf{r}, \mathbf{p}) = \frac{1}{2}\sum_{j\in\mathbb{Z}} \big(p_j^2 + \sum_{1\le k\le K} a_k |\sum_{0\le l<k} r_{j+l}|^2\big)$. The transformed Hamiltonian system reads as

$$(\dot{\mathbf{r}}, \dot{\mathbf{p}})^T = \mathcal{J}_{\mathrm{r}}\, \mathrm{d}\mathcal{H}_{\mathrm{r}}(\mathbf{r}, \mathbf{p})^T =: \mathcal{L}(\mathbf{r}, \mathbf{p})^T \tag{2.5a}$$

with

$$\mathcal{J}_{\mathrm{r}} := \begin{pmatrix} \mathbf{0} & \partial_1 - \mathbf{1} \\ \mathbf{1} - \partial_{-1} & \mathbf{0} \end{pmatrix}, \quad \mathrm{d}\mathcal{H}_{\mathrm{r}} = \begin{pmatrix} \sum_{|l|<K} \sum_{|l|<k\le K} (k - |l|) a_k \partial_l & \mathbf{0} \\ \mathbf{0} & \mathbf{1} \end{pmatrix}, \tag{2.5b}$$

and $(\partial_l \mathbf{z})_j = z_{j+l}$. The operator $\mathcal{J}_{\mathrm{r}}$ is a non-canonical Poisson structure arising from the push-forward of the Poisson tensor $\mathcal{J}_{\mathsf{can}}$, that is $\mathcal{J}_{\mathrm{r}} = \mathcal{T}\, \mathcal{J}_{\mathsf{can}}\, \mathcal{T}^*$ where $\mathcal{T}$ is the linear map defined by $(\mathbf{r}, \mathbf{p})^T = \mathcal{T}(\mathbf{x}, \mathbf{p})^T$.

Using the Fourier transform $\mathcal{F} : \ell^2(\mathbb{Z}, \mathbb{R}^2) \to L^2(\mathcal{S}^1, \mathbb{R}^2)$ defined by $\hat{z}(\theta) = \sum_{j\in\mathbb{Z}} z_j \mathrm{e}^{-\mathrm{i}j\theta}$, it is possible to solve (2.5) explicitly. Applying $\mathcal{F}$ leads to

$$\begin{pmatrix} \dot{\hat{\mathbf{r}}} \\ \dot{\hat{\mathbf{p}}} \end{pmatrix} = \begin{pmatrix} 0 & \mathrm{e}^{\mathrm{i}\theta} - 1 \\ 1 - \mathrm{e}^{-\mathrm{i}\theta} & 0 \end{pmatrix} \begin{pmatrix} \omega_{\mathrm{r}}^2(\theta) & 0 \\ 0 & 1 \end{pmatrix} \begin{pmatrix} \hat{\mathbf{r}} \\ \hat{\mathbf{p}} \end{pmatrix} , \tag{2.6}$$

where

$$\omega_{\mathrm{r}}^2(\theta) = \sum_{0<k\le K} k a_k + 2 \sum_{0<l\le K-1} \Big(\sum_{l<k\le K} (k-l) a_k \Big) \cos(l \cdot \theta). \tag{2.7}$$

Now, solving the linear system (2.6) we obtain the fundamental matrix $\hat{\mathbf{G}}_{\mathrm{r}}(t,\theta)$ and Green's function of our original problem is given by inverse Fourier transform, $\mathbf{G}(t) = \mathcal{F}^{-1}\{\hat{\mathbf{G}}_{\mathrm{r}}(t,\theta)\} = \frac{1}{2\pi} \int_{\mathcal{S}^1} \hat{\mathbf{G}}_{\mathrm{r}}(t,\theta)\,\mathrm{d}\theta$ with

$$G_j(t) = \frac{1}{2\pi} \int_{\mathcal{S}^1} \begin{pmatrix} \cos(\omega(\theta)t) & \frac{\mathrm{e}^{\mathrm{i}\theta}-1}{\omega(\theta)} \sin(\omega(\theta)t) \\ \frac{-\omega(\theta)}{\mathrm{e}^{-\mathrm{i}\theta}-1} \sin(\omega(\theta)t) & \cos(\omega(\theta)t) \end{pmatrix} \mathrm{e}^{\mathrm{i}j\theta}\,\mathrm{d}\theta , \qquad j \in \mathbb{Z} \tag{2.8}$$

Thus the long time behavior of solutions is determined by oscillatory integrals. Altogether we proved the following lemma.

Proposition 2.1 (Explicit solution):
Given some initial conditions $(\mathbf{r}^0, \mathbf{p}^0)^T \in \ell^2(\mathbb{Z}, \mathbb{R}^2)$, *the unique solution of* $(\dot{\mathbf{r}}, \dot{\mathbf{p}})^T = \mathcal{L}(\mathbf{r}, \mathbf{p})^T$ *defined in* (2.5) *is determined by*

$$(\mathbf{r}(t), \mathbf{p}(t))^T = \mathrm{e}^{\mathcal{L}t} (\mathbf{r}^0, \mathbf{p}^0)^T \tag{2.9a}$$

where $(\mathrm{e}^{\mathcal{L}t})_{t\in\mathbb{R}}$ *is a differentiable group of bounded operators on* $\ell^2(\mathbb{Z}, \mathbb{R}^2)$ *defined by*

$$\left(\mathrm{e}^{\mathcal{L}t}(\mathbf{r}, \mathbf{p})^T\right)_j = \sum_{k\in\mathbb{Z}} G_k(t) \cdot (r_{j-k}, p_{j-k})^T \qquad \text{for } j \in \mathbb{Z} \tag{2.9b}$$

with $G_j(t)$ *defined in* (2.8).

Now we want to characterize the dispersion relation more precisely. From now on we will assume that, additionally to the stability condition, the following non-degeneracy condition is satisfied,

$$\omega'(0) > 0 \qquad \text{and} \qquad \forall \hat{\theta} \in \mathcal{S}^1 : \omega''(\hat{\theta}) = 0 \implies \omega'''(\hat{\theta}) \neq 0 \tag{2.10}$$

These conditions as will be discussed below in connection with (2.16). In fact, these are not fundamental for the upcoming discussions, but a violation would lead to different decay rates and necessitate case-by-case analysis. Now we may highlight important properties of the dispersion relation ω in the following lemma.

Lemma 2.2:
For the dispersion relation defined by (2.2) *and* (2.4) *holds*

$$\omega(\theta) = 2\left|\sin\tfrac{\theta}{2}\right|\omega_{\mathrm{r}}(\theta) \tag{2.11}$$

with ω_{r} *given by* (2.7). *Furthermore, for* ω_{r} *holds*

$$\forall\,\theta \in \mathcal{S}^1 :\ \omega_{\mathrm{r}}(\theta) = \omega_{\mathrm{r}}(-\theta) = \omega_{\mathrm{r}}(\theta{+}2\pi) \tag{2.12a}$$

and, if additionally the stability and non-degeneracy conditions (2.3) *and* (2.10) *are satisfied, then*

$$\exists\, c_{\mathrm{r}} > 0\ \forall\,\theta \in \mathcal{S}^1 :\ \omega_{\mathrm{r}}(\theta) \geq c_{\mathrm{r}}\ . \tag{2.12b}$$

Proof. Since the linear equation (2.1) is invariant under the transformation $(\partial_1{-}\mathbf{1})$, the first statement follows from $\ddot{\hat{\mathbf{r}}} = -\Lambda(\theta)\hat{\mathbf{r}}$ and the fact that (2.6) implies $\ddot{\hat{\mathbf{r}}} = 2(\cos\theta - 1)\omega_r(\theta)^2\hat{\mathbf{r}}$. The symmetry and periodicity ω_{r} is obvious from the explicit formula (2.7). To see $c_{\mathrm{r}} > 0$ note first that, in view of (2.3), it is sufficient to check $\omega_{\mathrm{r}}(0) \neq 0$. But this follows from $\omega'(0) = \omega_{\mathrm{r}}(0)$ and (2.10). □

Note that the first factor of ω is due to the transformation of the Poisson tensor and arises independently of the actual interaction of the particles. For that reason we carried out the transformation in detail.

Finally, we state a second equivalent representation of Green's function which will be used below. It is obtained by rewriting the dispersion relation,

$$\tilde{\omega}(\theta) = 2\sin\tfrac{\theta}{2}\,\omega_{\mathrm{r}}(\theta) \tag{2.13}$$

and using the symmetry of ω_{r}, namely

$$G(t,c) = \frac{1}{2\pi}\int\limits_{\mathcal{S}^1}\begin{pmatrix} \mathrm{e}^{\mathrm{i}t\phi_\pm(2\theta,c)} & \pm\frac{\mathrm{e}^{\mathrm{i}\theta}}{\omega_{\mathrm{r}}(2\theta)}\mathrm{e}^{\mathrm{i}t\phi_\pm(2\theta,c)} \\ \pm\frac{\omega_{\mathrm{r}}(2\theta)}{\mathrm{e}^{\mathrm{i}\theta}}\mathrm{e}^{\mathrm{i}t\phi_\pm(2\theta,c)} & \mathrm{e}^{\mathrm{i}t\phi_\pm(2\theta,c)}\end{pmatrix}\mathrm{d}\theta$$
$$\text{where}\quad \phi_\pm(\theta,c) = \theta c \pm \tilde{\omega}(\theta) \quad\text{and}\quad c = \frac{j}{t}\ . \tag{2.14}$$

The proof as well as further useful representations of G are given in Appendix A. Note that $\tilde{\omega}$ now is 4π-periodic. The new variable $c \in \mathbb{R}$ characterizes the rays $j = ct$ and refers to the the group velocity $c_{\mathrm{gr}}(\theta) = \pm\omega'(\theta)$. In view of the fact that $\theta \mapsto c_{\mathrm{gr}}(\theta)$ is not injective, it is useful to define

$$\mathbf{\Theta}(c) := \{\theta \in \mathcal{S}^1\,|\,\omega'(\theta) = c\}\ .$$

Note that, although we will sometimes consider c being a continuous variable, it is to be evaluated in j/t. This is crucial in view of the fact that $\theta \mapsto \mathrm{e}^{\mathrm{i}t\phi_\pm(2\theta,c)}$ remains 2π-periodic for that choice.

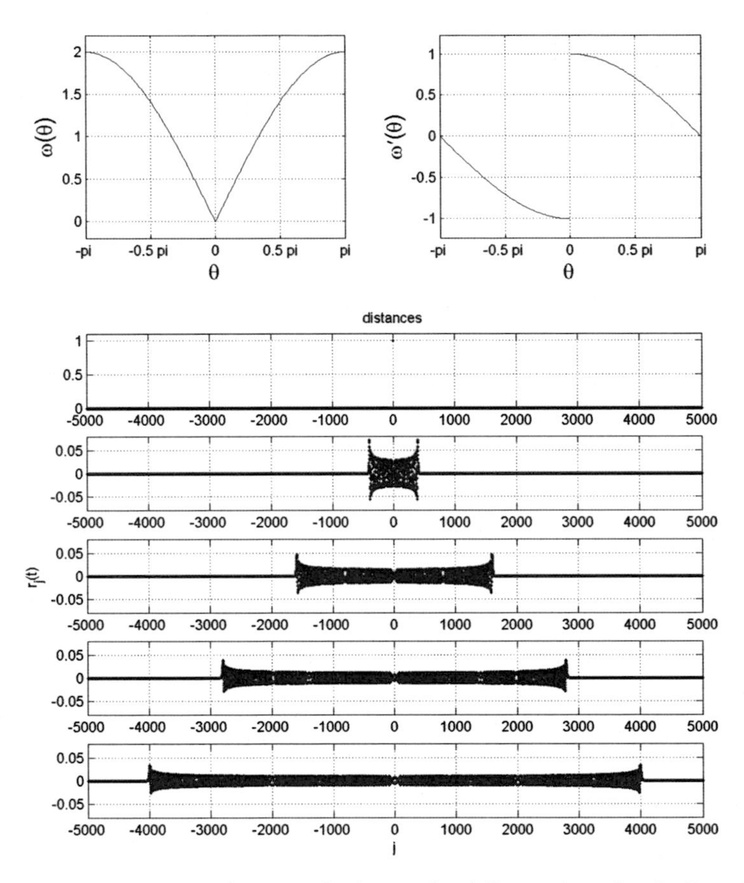

Figure 2.1: Dispersion relation and solutions for different times for the linearized FPU, i.e. $\omega(\theta) = 2|\sin\frac{\theta}{2}|$.

2.2 Critical wave numbers, dispersive decay and finite sonic velocity

The asymptotic behavior of solutions of (2.5) is dominated by dispersion occurring in such lattices systems. A typical consequence in this context is the decay of solutions to localized initial conditions: Consider the group $(e^{\mathcal{L}t})_{t\in\mathbb{R}}$ in (2.9) and assume that the dispersion relation ω satisfies the stability condition (2.3) and the non-degeneracy condition (2.10). Then, for $p \in [2, 4) \cup (4, \infty]$ there exists C_p such

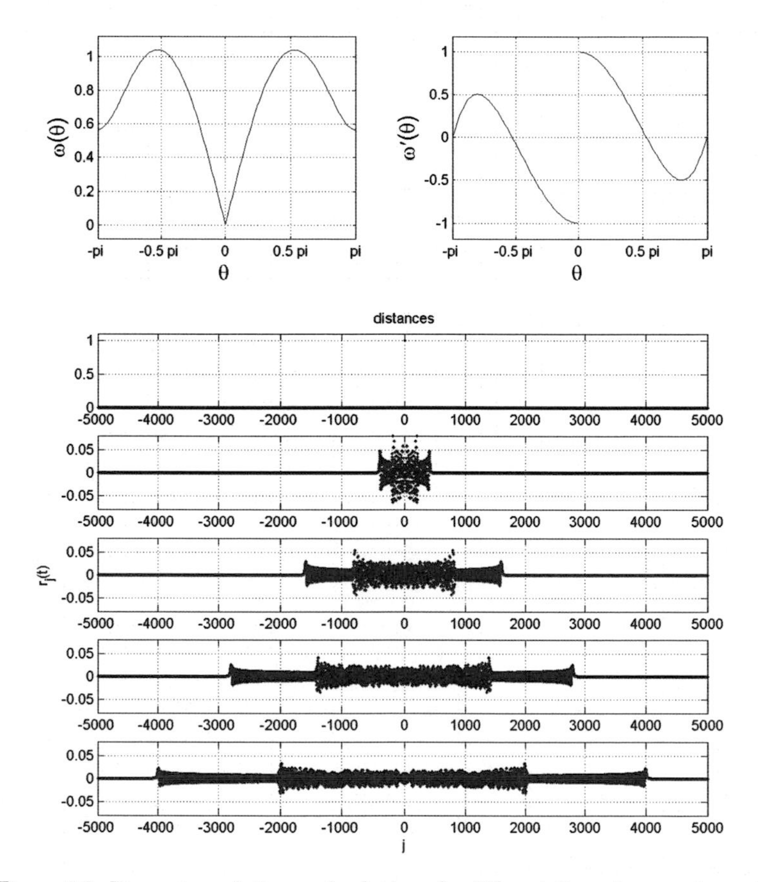

Figure 2.2: Dispersion relation and solutions for different times for a system with second nearest-neighbor interaction with $a_1 = 0.08$ and $a_2 = 0.23$. The coefficients are chosen such that the group velocities are approximately 1 and 1/2.

that, for all $t \geq 0$, we have

$$\|e^{\mathcal{L}t}\|_{\ell^1,\ell^p} \leq \frac{C_p}{(1+t)^{\alpha_p}}, \quad \text{where} \quad \alpha_p = \begin{cases} \dfrac{p-2}{2p} & \text{for } p \in [2,4), \\ \dfrac{p-1}{3p} & \text{for } p \in (4,\infty]. \end{cases} \tag{2.15}$$

Here we skipped the case $p = 4$ where an additional logarithmic correction term occurs, see [MP10] for details and the proof. Like in PDE theory, this decay estimate carries over to nonlinear equations if the nonlinearity is weak and the initial conditions are sufficiently small, see again [MP10] and also [SK05, GHM06].

To obtain these decay rates α_p, which are better than those one gets by standard Riesz Thorin interpolation based on $\alpha_2 = 0$ and $\alpha_\infty = 1/3$, the local decay of $(e^{\mathcal{L}t})_{t\in\mathbb{R}}$ needs to be estimated carefully. We will from now on focus on this local behavior. In the remaining section we summarize important results concerning upper bounds of solutions. In Section 3 and 4, respectively, we will derive uniform asymptotic expansions for the solutions. In this context we will see that these results are optimal - not only in term of the decay rates, but also in terms of $c = j/t$.

The dispersive decay relies on the fact that the behavior of solutions to (2.5) is determined by oscillatory integrals of the form

$$g(t,c) = \int_{\mathcal{S}^1} A(\theta) e^{it\phi(\theta,c)} \, d\theta \tag{2.16}$$

with $A(\theta)$ standing for 1, $1/\omega_r(\theta)$ or $\omega_r(\theta)$. According to Lemma 2.2, A is real-analytic in any case. Thus, oscillations with wave numbers θ travel along rays $j = c_{gr}(\theta)t$, where the group velocity is defined by the relation $\partial_\theta\phi(\theta, c_{gr}(\theta)) = 0$, i.e. $c_{gr}(\theta) = \pm\tilde{\omega}'(\theta)$. The method of stationary phase, see e.g. [Ste93, Won89, Whi74] indicates that the decay along these rays is like $t^{-1/2}$ if $\partial_\theta^2\phi(\theta, c) = \tilde{\omega}''(\theta) \neq 0$ and like $t^{-1/3}$ for $\tilde{\omega}''(\theta) = 0$. Weaker decay rates, for instance $t^{-1/4}$, are excluded by the non-degeneracy condition (2.10). We define the set of critical wave numbers $\mathbf{\Theta}_{cr}$ and the maximal wave speed c_{max} via

$$\mathbf{\Theta}_{cr} := \left\{ \hat{\theta} \in \mathcal{S}^1 \mid \tilde{\omega}''(\hat{\theta}) = 0 \right\} \quad \text{and} \quad c_{max} := \max \left\{ \, |\tilde{\omega}'(\theta)| \, \middle| \, \theta \in \mathcal{S}^1 \right\}.$$

Note that $0 \in \mathbf{\Theta}_{cr}$ and, since K in (2.2) is finite, the set $\mathbf{\Theta}_{cr}$ is discrete. In this paper we in particular aim to understand the crossover between the two different decay rates, i.e. the behavior for $c \in \mathcal{C}_{cr}(\varepsilon)$ with

$$\mathcal{C}_{cr}(\varepsilon) := \bigcup_{\hat{\theta}\in\mathbf{\Theta}_{cr}} \left[c_{gr}(\hat{\theta}) - \varepsilon, c_{gr}(\hat{\theta}) + \varepsilon \right] \subset \mathbb{R}.$$

In the upcoming sections two other sets will be relevant. For the sake of completeness we define these already here.

$$\mathbf{\Theta}_{cr}(\delta) := \bigcup_{\hat{\theta}\in\mathbf{\Theta}_{cr}} \overline{\mathcal{U}_\delta(\hat{\theta})} \subset \mathcal{S}^1, \qquad \mathbf{\Theta}_{cr}(c,\varepsilon) := \left\{ \hat{\theta} \in \mathbf{\Theta}_{cr} \, \middle| \, \left| c_{gr}(\hat{\theta}) - c \right| \leq \varepsilon \right\}$$

The first is analogous to $\mathcal{C}_{cr}(\varepsilon)$, but in terms of the wave numbers and the $\mathbf{\Theta}_{cr}(c,\varepsilon)$ characterizes critical wave numbers corresponding to group velocities in a neighborhood of a given group velocity c.

In Figure 2.1 and 2.2 we plot two dispersion relations and associated solutions $r_j(t)$ for different times to display the influence of the critical wave numbers

$\hat{\theta}_j \in \mathbf{\Theta}_{\rm cr}$. As a consequence of the Galilean invariance the wave number $\theta = 0$ is outstanding in two respects (cf. (2.11)). First, as already mentioned above, $0 \in \mathbf{\Theta}_{\rm cr}$ such that there are always wave fronts traveling with speed $c = \pm\tilde{\omega}'(0)$ and second, since $\tilde{\omega}(0) = 0$, these wave front are monotone. We will prove this monotonicity of the front in Section 4. The latter holds for all $t > 0$ (not only in the limit $t \to \infty$) if $\tilde{\omega}'(0) = \max_{\theta \in S^1} \tilde{\omega}'(\theta) := c_{\max}$, i.e. the sonic velocity, which is not satisfied in general for $K > 1$.

The tool to derive upper bound on (2.16) is van der Corput's lemma, see e.g. [Ste93]. It states that if $\left|\phi^{(k)}(\theta)\right| \geq \lambda > 0$ for $\theta \in (\underline{\theta}, \overline{\theta})$ where either $k \geq 2$, or $k = 1$ and ϕ' is monotonic, then

$$\left| \int_{\underline{\theta}}^{\overline{\theta}} A(\theta) \mathrm{e}^{\mathrm{i}t\phi(\theta)} \,\mathrm{d}\theta \right| \leq C(k, A)\, (\lambda t)^{-1/k} \tag{2.17}$$

with $C(k, A) = (5{\cdot}2^{k-1}{-}2) \left(\max_{\theta \in [\underline{\theta}, \overline{\theta}]} |A(\theta)| + \int_{\underline{\theta}}^{\overline{\theta}} |A(\theta)| \mathrm{d}\theta \right)$. Based on this a global bound on $g(t, c)$ is straightforward, cf. [MP10, Lemma 3.5]: There exists a constant $C(\tilde{\omega}, A) > 0$ depending on the dispersion relation $\tilde{\omega}$ and A such that

$$\forall t \geq 0,\ c \in \mathbb{R}: \quad |g(t, c)| \leq \frac{C(\tilde{\omega}, A)}{(1+t)^{1/3}}. \tag{2.18}$$

Similarly, it is possible to obtain an upper bound of order $t^{-1/2}$ for c bounded away from group velocities corresponding to critical wave numbers, i.e. for $c \notin \mathcal{C}_{\rm cr}(\varepsilon)$ with $\varepsilon > 0$. But a careful application of van der Corput's lemma gives an upper bound for the singularity occurring as $c \to \tilde{\omega}'(\hat{\theta})$, $\hat{\theta} \in \mathbf{\Theta}_{\rm cr}$. Namely, according to [MP10, Lemma 3.6] we can choose $\varepsilon = t^{-2/3}$, and there exists again a constant $\tilde{C}(\tilde{\omega}, A) > 0$ such that

$$\begin{aligned} &\forall t \geq 0\ \forall c \in \mathbb{R} \setminus \mathcal{C}_{\rm cr}(t^{-2/3}): \\ &|g(t, c)| \leq \frac{\tilde{C}(\tilde{\omega}, A)}{(1+t)^{1/2}} \left(1 + \sum_{\hat{\theta} \in \mathbf{\Theta}_{\rm cr}} \frac{1}{|\tilde{\omega}'(\hat{\theta})^2 - c^2|^{1/4}} \right). \end{aligned} \tag{2.19}$$

Combining both upper bounds suggests that the wave fronts with decay $t^{-1/3}$ expand like $t^{1/3}$ (in terms of $r_j(t)$). In Section 3 we will see that this indeed holds and furthermore that (2.19) is sharp.

The third important decay estimate concerns the exponential decay of $g(t, c)$ for $|c| > c_{\max}$. Thus, ignoring exponentially small tails, this is equivalent to a finite sonic velocity. Although we suppose that the result is known we give the statement and the full proof since we were not able to find a citable reference. Only in [Fri03] there is a similar statement for the case of pure nearest-neighbor interaction (with a different proof), but unfortunately this work is unpublished. For this result we choose $\gamma > 0$ such that $\omega_{\rm r}$ can be holomorphically extended on the complete stripe $\{\theta \in \mathbb{C} \,|\, 0 \leq \mathfrak{Im}\, \theta \leq \gamma\}$ where $\omega_{\rm r}(\theta) > 0$ still holds.

Proposition 2.3:
Consider the Green's function defined in (2.9) with dispersion relation ω satisfying the stability and non-degeneracy condition (2.3) and (2.10), respectively. Then, for all $t>0$ and all $c=\frac{j}{t}\in\mathbb{R}\setminus[-c_{\max},c_{\max}]$ holds

$$|G(t,c)| \le \begin{pmatrix} 1 & \mathrm{e}^{-\operatorname{sign}(c)\alpha_*(c)/2}/\underline{\omega}_*(c) \\ \mathrm{e}^{\operatorname{sign}(c)\alpha_*(c)/2}\,\overline{\omega}_*(c) & 1 \end{pmatrix} \mathrm{e}^{-\kappa(c)t} \tag{2.20a}$$

with

$$\kappa(c) := \max_{\alpha\in[0,\gamma]}\left(\alpha|c| - \max_{\theta\in[-\pi,\pi]} \mathfrak{Im}\,\tilde{\omega}(2\theta+\mathrm{i}\alpha)\right) > 0, \tag{2.20b}$$

$\underline{\omega}_*(c) := \min_{\theta\in[-\pi,\pi]}|\omega_{\mathrm{r}}(2\theta+\mathrm{i}\alpha_*(c))|$ *and* $\overline{\omega}_*(c) := \max_{\theta\in[-\pi,\pi]}|\omega_{\mathrm{r}}(2\theta+\mathrm{i}\alpha_*(c))|$, *where* $\alpha_*(c)$ *maximizes the term defining* $\kappa(c)$.

Proof. According to (2.14) the components of $G_j(t)$ are given by

$$g_\pm(t,c) = \frac{1}{2\pi}\int_{-\pi}^{\pi} A(\theta)\mathrm{e}^{\mathrm{i}t(2\theta c\pm\tilde{\omega}(2\theta))}\,\mathrm{d}\theta \quad \text{with} \quad c=\frac{j}{t},$$

in which $A(\theta)=1, \frac{\mathrm{e}^{\mathrm{i}\theta}}{\omega_{\mathrm{r}}(2\theta)}$ or $\omega_{\mathrm{r}}(2\theta)\mathrm{e}^{-\mathrm{i}\theta}$. In all cases $A(\theta)$ is analytic and 2π-periodic.

To prove the statement for $c>c_{\max}$ we consider $g_+(t,c)$ and continue $\tilde{\omega}$ and A analytically to $[-\pi,\pi]+\mathrm{i}\,[0,\gamma]$. Applying Cauchy's integral theorem we get for $\alpha>0$

$$\int_{-\pi}^{\pi} A(\theta)\mathrm{e}^{\mathrm{i}t(2\theta c+\tilde{\omega}(2\theta))}\,\mathrm{d}\theta = \left(\int_{-\pi}^{-\pi+\mathrm{i}\alpha} + \int_{-\pi+\mathrm{i}\alpha}^{\pi+\mathrm{i}\alpha} + \int_{\pi+\mathrm{i}\alpha}^{\pi}\right) A(\theta)\mathrm{e}^{\mathrm{i}t(2\theta c+\tilde{\omega}(2\theta))}\,\mathrm{d}\theta\,.$$

Exploiting the symmetry of $A(\theta)$ we find

$$\begin{aligned} &\int_{-\pi}^{-\pi+\mathrm{i}\alpha} A(\theta)\mathrm{e}^{\mathrm{i}t(2\theta c+\tilde{\omega}(2\theta))}\,\mathrm{d}\theta + \int_{\pi+\mathrm{i}\alpha}^{\pi} A(\theta)\mathrm{e}^{\mathrm{i}t(2\theta c+\tilde{\omega}(2\theta))}\,\mathrm{d}\theta \\ &\qquad = 2\int_0^{\alpha} A(\pi+\mathrm{i}\alpha)\sin(2\pi ct)\mathrm{e}^{(-2yc+\mathrm{i}\tilde{\omega}(2\pi+2\mathrm{i}y))t}\,\mathrm{d}y = 0 \end{aligned}$$

since $c=\frac{j}{t}$. For the remaining integral holds

$$\begin{aligned} \left|\int_{-\pi+\mathrm{i}\alpha}^{\pi+\mathrm{i}\alpha} A(\theta)\mathrm{e}^{\mathrm{i}t(2\theta c+\tilde{\omega}(2\theta))}\,\mathrm{d}\theta\right| &= \mathrm{e}^{-2\alpha ct}\left|\int_{-\pi}^{\pi} A(\theta+\mathrm{i}\alpha)\mathrm{e}^{\mathrm{i}t(2\theta c+\tilde{\omega}(2\theta+2\mathrm{i}\alpha))}\,\mathrm{d}\theta\right| \\ &\le \mathrm{e}^{-2\alpha ct}\cdot 2\pi \max_{\theta\in[-\pi,\pi]}|A(\theta+\mathrm{i}\alpha)|\,\mathrm{e}^{t\max_{\theta\in[-\pi,\pi]}\mathfrak{Im}\,\tilde{\omega}(2\theta+2\mathrm{i}\alpha)}\,. \end{aligned}$$

Now we choose α such that the overall exponent is minimized and set $\alpha_*=2\alpha$ to obtain

$$|g_+(t,c)| \le \max_{\theta\in[-\pi,\pi]}\left|A\left(\theta+\mathrm{i}\frac{\alpha_*}{2}\right)\right|\,\mathrm{e}^{-\kappa(c)t}$$

with $\kappa(c)$ defined in (2.20b) which proves (2.20a). To see $\kappa(c)>0$ for $c>c_{\max}$ note that $\max_{\theta\in[-\pi,\pi]}\mathfrak{Im}\,\tilde{\omega}(2\theta+\mathrm{i}\alpha) = c_{\max}\alpha+\mathcal{O}(\alpha^3)$ as $\alpha\to 0$.

To prove the statements for $c < -c_{\max}$ we consider the representation

$$g_-(t,c) = \frac{1}{2\pi}\int_{-\pi}^{\pi} A(\theta)\mathrm{e}^{\mathrm{i}t(2\theta(-c)+\tilde{\omega}(2\theta))}\,\mathrm{d}\theta$$

with $A(\theta) = 1, \frac{\mathrm{e}^{-\mathrm{i}\theta}}{\omega_{\mathrm{r}}(2\theta)}$ or $\omega_{\mathrm{r}}(2\theta)\mathrm{e}^{\mathrm{i}\theta}$. Thus, the proof follows by the same arguments. □

Solving condition (2.20b) locally, the decay of solutions in terms of $c - c_{\max}$ near the wave fronts becomes evident: Consider $0 \le c - c_{\max} \le \varepsilon$ and $0 < \alpha \le \alpha_0$ for $\varepsilon, \alpha_0 > 0$ sufficiently small. Then there exists $B = B(\tilde{\omega}, \alpha_0) > 0$ such that

$$\begin{aligned}\kappa(c) &\ge c\alpha - c_{\max}\alpha - B(\tilde{\omega},\alpha_0)\alpha^3\\ &= (1-\xi)(c-c_{\max})\alpha + (1-\xi)(c-c_{\max})\alpha - B(\tilde{\omega},\alpha_0)\alpha^3\end{aligned}$$

with $0 < \xi < 1$. Now we claim $(1-\xi)(c-c_{\max})\alpha - B(\tilde{\omega},\alpha_0)\alpha^3 = 0$. This implies $\alpha = \sqrt{\frac{\varepsilon}{B(\tilde{\omega},\alpha_0)}(c-c_{\max})}$. Thus, for $0 \le c - c_{\max} \le \varepsilon$ with $\varepsilon > 0$ sufficiently small there exists $\tilde{\kappa}$ such that

$$\forall t \ge 0 : \quad |G(t,c)| \le C_{\omega,A}\mathrm{e}^{-\tilde{\kappa}(c-c_{\max})^{-3/2}t}\,. \tag{2.21}$$

3 Uniform asymptotic behavior

In this section we first state the main results. Namely, the uniform asymptotic expansions of solutions of (2.5) near wave fronts where the decay rate is $\sim t^{-1/3}$ and in the inner regions where we have a decay in time $\sim t^{-1/2}$. Here the leading order behavior as well as the Airy-like wave fronts are well known. Our contribution is to make the dependency on $c - c_{\mathrm{gr}}(\hat{\theta})$ for $\hat{\theta} \in \Theta_{\mathrm{cr}}$ more explicit. Afterwards we discuss the results, in particular the crossover between the different scales and give illustrating examples. The actual proofs are shifted out to Section 4.

3.1 Asymptotic expansions

Now we are in place to state the main results. The first concerns the non-degenerate case, namely $c \in [-c_{\max}, c_{\max}] \setminus \mathcal{C}_{\mathrm{cr}}(\varepsilon)$ for some $\varepsilon > 0$. We emphasize that the obtain we obtain will blow up, since the expansion becomes singular as $c \longrightarrow c_{\mathrm{gr}}(\hat{\theta})$ for $\hat{\theta} \in \Theta_{\mathrm{cr}}$. It would be interesting to keep track of the blowup behavior to optimize the overall error. However we are content with some, not necessarily optimal bound.

Theorem 3.1 (Asymptotic behavior in the non-degenerate case):
Consider Green's function of the Hamiltonian system (2.5) *and assume that the corresponding dispersion relation* ω *defined in* (2.4) *satisfies the non-degeneracy*

condition (2.10). *Then, for all* $\varepsilon > 0$ *there exists a constant* $C_{\rm nd}(\varepsilon,\omega) > 0$ *such that for all* $c \in [-c_{\max}, c_{\max}] \setminus C_{\rm cr}(\varepsilon)$ *and* $t > 0$ *we have the estimate*

$$|G(t,c) - \mathcal{G}_{\rm non}(t,c)| \leq C_{\rm nd}(\varepsilon,\omega)\, t^{-1} \quad \textit{with} \tag{3.1a}$$

$$\mathcal{G}_{\rm non}(t,c) = \sum_{\theta\in\theta(c)} \frac{1}{\sqrt{8\pi|\omega''(\theta)|}} \begin{pmatrix} 1 & \frac{1}{\omega_{\rm r}(\theta)} \\ \omega_{\rm r}(\theta) & 1 \end{pmatrix} \cos\left[(\omega(\theta)+c\theta)t + \operatorname{sign}\omega''(\theta)\tfrac{\pi}{4}\right] t^{-1/2}\,. \tag{3.1b}$$

Indeed, for the constant in (3.1a) we will show the upper bound

$$C_{\rm in}(\varepsilon,\omega) = C_1(\omega) + \frac{C_2(\omega)}{\varepsilon^{5/2}} \tag{3.2}$$

as $\varepsilon \to 0$. Note furthermore that the coefficient in (3.1b) becomes unbounded in that case. According to Lemma 4.1 holds $\sqrt{|\omega''(\theta)|} \sim \varepsilon^{1/4}$ as $\theta \to \hat{\theta} \in \Theta_{\rm cr}$.

Concerning the second result, the challenge is that the dispersion relation degenerates. Thus, in the general case the implicit function theorem is not sufficient. Here a suitable version of Weierstrass preparation theorem is necessary. In this context a classical statement is the following.

Theorem 3.2 (Method of stationary phase, cf. [Hör90, 7.7.18]):
Let ϕ *be a real-valued* C^∞ *function of* $(\theta, y) \in \mathbb{R}^{1+n}$ *near* 0 *such that*

$$\phi(0,0) = \partial_\theta\phi(0,0) = \partial_\theta^2\phi(0,0) = 0 \quad \textit{and} \quad \partial_\theta^3\phi(0,0) \neq 0\,.$$

Then there exist C^∞ *real valued functions* $a(y)$ *and* $b(y)$ *near* 0 *such that* $a(0) = b(0) = 0$ *and*

$$\begin{aligned}\int A(\theta,y)\mathrm{e}^{\mathrm{i}t\phi(\theta,y)}\,\mathrm{d}\theta \sim \mathrm{e}^{\mathrm{i}tb(y)}\Big(&\mathrm{Ai}\big(a(y)t^{2/3}\big)\,t^{-1/3}\sum_{0\leq j\leq\infty} u_{0,j}(y)t^{-j} \\ &+ \mathrm{Ai}'\big(a(y)t^{2/3}\big)\,t^{-2/3}\sum_{0\leq j\leq\infty} u_{1,j}(y)t^{-j}\Big)\,,\end{aligned} \tag{3.3}$$

provided that $A \in C_0^\infty$ *and* $\operatorname{supp} A$ *is sufficiently close to* 0. *Here* $u_{0,j}$, $u_{1,j} \in C_0^\infty$.

Here $\mathrm{Ai}(\cdot)$ denotes Airy's function defined by $\mathrm{Ai}(z) = \frac{1}{2}\int_{\mathbb{R}} \mathrm{e}^{i(u^3/3+zu)}\mathrm{d}z$, see for instance [Olv74] for basic properties. Figure 3.1 show wave fronts together with the correspondingly scale Airy function.

There remain two questions. First, the result does neither state the functions a, b, $u_{0,j}$ and $u_{1,j}$ nor the error bounds in terms explicitly. Second, these functions are not determined uniquely, i.e. the theorem does not state that the asymptotic series (3.3) is unique. We will make these functions and the error estimates for this expansion more explicit in Section 4.3. For this we give a short summary on a suitable version of Weierstrass' preparation theorem and explain that the functions

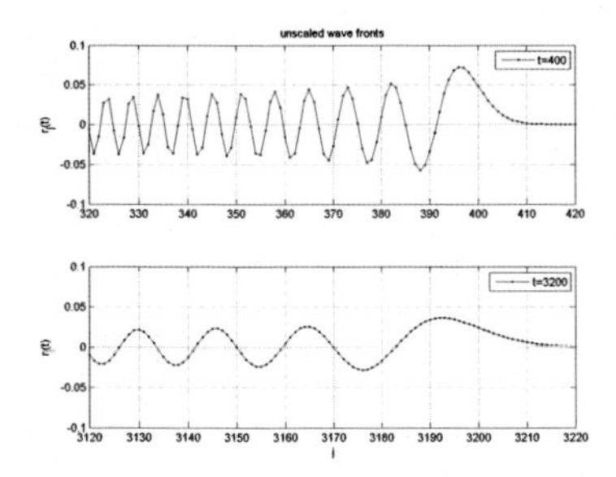

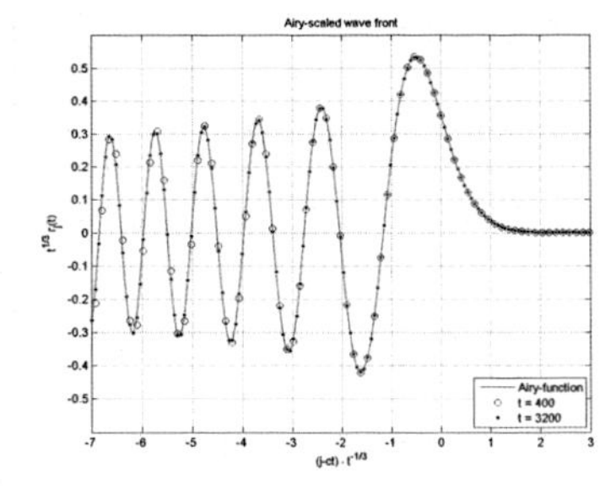

Figure 3.1: Wave fronts for the linearized FPU chain in comparison with the corresponding Airy function approximation.

a and b can be made more explicit. This allows us to estimate the remainder terms explicitly.

For the following result we recall that that for a degenerate wave number $\theta_{\rm cr}$ there may be other non-degenerate wave numbers θ having the same group velocity, namely $c_{\rm gr}(\theta_{\rm cr}) = c_{\rm gr}(\theta)$. Thus, an Airy-type degenerate behavior may be superimposed by a non-degenerate harmonic wave train. This is expressed in the following result by the additive structure $\mathcal{G}_{\rm dg} + \mathcal{G}_0$ and can be see in the example displayed in Figure 2.2.

Theorem 3.3 (Asymptotic behavior near degenerate points):
Consider Green's function of the Hamiltonian system (2.5) *and assume that the corresponding dispersion relation* ω *defined in* (2.4) *satisfies the non-degeneracy condition* (2.10). *Then, there exists* $\varepsilon_0 = \varepsilon_0(\tilde{\omega}) > 0$, $\delta_0 = \delta_0(\tilde{\omega}) > 0$ *and* $C_{\rm dg}(\tilde{\omega}) > 0$ *such that for all* $c \in \mathcal{C}_{\rm cr}(\varepsilon_0)$

$$\big|G(t,c) - \mathcal{G}_{\rm dg}(t,c) - \mathcal{G}_0(t,c)\big| \leq C_{\rm dg}(\omega)t^{-1} \tag{3.4a}$$

with

$$\begin{aligned}\mathcal{G}_{\rm dg}(t,c) = \sum_{\hat{\theta}\in\Theta_{\rm cr}(c,\varepsilon_0)} \Big(&\mathbb{A}_{\hat{\theta}}(c)\cos\Big(\big[\omega(\hat{\theta}) - \hat{c}\hat{\theta} - b_{\hat{\theta}}(c)\big]t\Big)\,\mathrm{Ai}\left(a_{\hat{\theta}}(c)t^{2/3}\right)t^{-1/3} \\ &- \mathbb{B}_{\hat{\theta}}(c)\sin\Big(\big[\omega(\hat{\theta}) - \hat{c}\hat{\theta} - b_{\hat{\theta}}(c)\big]t\Big)\,\mathrm{Ai}'\left(a_{\hat{\theta}}(c)t^{2/3}\right)t^{-2/3}\Big),\end{aligned} \tag{3.4b}$$

where $\hat{c} = \omega'(\hat{\theta})$, *and*

$$\begin{aligned}\mathcal{G}_0(t,c) = \sum_{\substack{\theta\in\Theta(c)\ \mathit{and}\\ \mathrm{dist}(\theta,\Theta_{\rm cr}(c,\varepsilon_0))>\delta_0}} &\frac{1}{\sqrt{8\pi|\omega''(\theta)|}}\begin{pmatrix} 1 & \frac{1}{\omega_{\rm r}(\theta)} \\ \omega_{\rm r}(\theta) & 1\end{pmatrix} \\ &\cdot\cos\left[(\omega(\theta) - c\theta)t + \operatorname{sign}\omega''(\theta)\tfrac{\pi}{4}\right]t^{-1/2}\,.\end{aligned} \tag{3.4c}$$

The scalar-valued functions $a_{\hat{\theta}}$, $b_{\hat{\theta}}$ as well as the $\mathbb{R}^{2\times 2}$-valued functions $\mathbb{A}_{\hat{\theta}}$ and $\mathbb{B}_{\hat{\theta}}$ are real-analytic on $[\hat{c}-\varepsilon_0, \hat{c}+\varepsilon_0]$. Furthermore, for $(c-\omega'(\hat{\theta}))\omega'''(\hat{\theta}) > 0$ we have

$$a_{\hat{\theta}}(c) = \left(\tfrac{3}{4}[\varsigma_-(c) - \varsigma_+(c)]\right)^{2/3} = -\sqrt[3]{\tfrac{2}{\tilde{\omega}'''(\hat{\theta})}}\,(c-\hat{c}) + \mathcal{O}\left((c-\hat{c})^2\right) \qquad (3.5)$$

$$b_{\hat{\theta}}(c) = \tilde{\omega}(\hat{\theta}) - \hat{c}\hat{\theta} - \tfrac{1}{2}[\varsigma_-(c) + \varsigma_+(c)] = 2\hat{\theta}(c-\hat{c}) + \mathcal{O}\left((c-\hat{c})^2\right)\ , \qquad (3.6)$$

where $\varsigma_\pm(c) := \tilde{\omega}(\theta_\pm(c)) - c\theta_\pm(c)$ and $\theta_+(c), \theta_-(c) \in \mathcal{U}_{\delta_0}(\hat{\theta})$ are the two wave numbers such that $c = \tilde{\omega}'(\theta_\pm(c))$ and $\theta_-(c) < \theta_+(c)$. Moreover, we have

$$\mathbb{A}_{\hat{\theta}}(c) = \frac{1}{\sqrt[3]{4|\tilde{\omega}'''(\hat{\theta})|}}\begin{pmatrix} 1 & \frac{1}{\omega_{\mathrm{r}}(\hat{\theta})} \\ \omega_{\mathrm{r}}(\hat{\theta}) & 1 \end{pmatrix} + \mathcal{O}(c-\hat{c})\ . \qquad (3.7)$$

Note that the error bound $C_{\mathrm{dg}}(\omega)$ depends on ω only. If c is near the sonic wave speed $c_{\max}$ the non-degenerate part $\mathcal{G}_0$ vanishes and the sum in (3.4b) reduces to one term (or in the degenerated case of several wave numbers traveling with the same sonic wave speed to just as many terms). Furthermore in case $\hat{\theta} = 0 \bmod \pi$ we have $b_{\hat{\theta}} \equiv 0$.

3.2 Full approximation and crossover

We are now in the position to define the full expansion that gives rise to the uniform approximation result stated in Theorem 1.1. We simply choose the ε_0 according to Theorem 3.3 and define $\mathcal{G}^{\mathrm{expan}}$ via

$$\mathcal{G}^{\mathrm{expan}}(t,c) = \begin{cases} \mathcal{G}_{\mathrm{non}}(t,c) & \text{for } c \in [-c_{\max}, c_{\max}] \setminus \mathcal{C}_{\mathrm{cr}}(\varepsilon_0), \\ \mathcal{G}_{\mathrm{deg}}(t,c) + \mathcal{G}_0(t,c) & \text{for } c \in \mathcal{C}_{\mathrm{cr}}(\varepsilon_0), \\ 0 & \text{for } |c| > c_{\max} + \varepsilon_0. \end{cases} \qquad (3.8)$$

Combining Proposition 2.3 and the Theorems 3.1 and 3.3 we have established the main Theorem 1.1 on the uniform approximation of the exact Green's function by $\mathcal{G}^{\mathrm{expan}}$ up to a uniform error of order $\mathcal{O}(1/t)$.

Now we discuss the crossover between the three different regions of definition for the function $\mathcal{G}^{\mathrm{expan}}$. We first look at the expansions for $|c| > c_{\max}$. For all such c we have exponential decay for $t \to \infty$, by Proposition 2.3. Thus, replacing this function by 0 will certainly keep the error order $\mathcal{O}(1/t)$.

Second we look at the overlap between the degenerate and the non-degenerate case inside $(-c_{\max}, c_{\max})$. Certainly we can make ε_0 for the degenerate as large as possible and make ε for the non-degenerate much smaller. Then we have an overlap of the two domains where the expansions are valid. Since both approximations decay slower that $1/t$, the theory can only be consistent if the error between the two approximations is at most $\mathcal{O}(1/t)$. To understand the crossover between these to regimes explicitly, we consider $\hat{\theta} \in \mathbf{\Theta}_{\rm cr}$ with $\tilde{\omega}'''(\hat{\theta}) > 0$, i.e. a wave front traveling to $j \to +\infty$ as $t \to \infty$. The leading order term of (3.4b) is

$$g_{\rm deg}(t,c) = \mathbb{A}_{\hat{\theta}}(c) \cdot \cos\Big(\big[\omega(\hat{\theta}) - \hat{c}\hat{\theta} - b_{\hat{\theta}}(c)\big]t\Big) \cdot \mathrm{Ai}\left(a_{\hat{\theta}}(c)t^{2/3}\right)t^{-1/3} .$$

We fix $c \in \tilde{\omega}'(S^1)$ such that $-\varepsilon_0 < c - \hat{c} < 0$ and aim to determine the behavior as $t \to \infty$. Recall the asymptotic behavior of Airy's function,

$$\mathrm{Ai}(z) \sim \frac{1}{\sqrt{\pi}} z^{-1/4}\Big(\cos\left(\tfrac{2}{3}z^{3/2} - \tfrac{\pi}{4}\right) + \mathcal{O}(z^{-3/2})\Big) \qquad \text{as} \quad z \to -\infty .$$

Now, $z = a_{\hat{\theta}}(c)t^{2/3}$ and (3.5) imply $\frac{2}{3}z^{3/2} = \frac{1}{2}\big[\varsigma_-(c) - \varsigma_+(c)\big]$. With (3.6) we find

$$\cos\Big(\big[\omega(\hat{\theta}) - \hat{c}\hat{\theta} - b_{\hat{\theta}}(c)\big]t\Big)\cos\left(\tfrac{2}{3}z^{3/2} - \tfrac{\pi}{4}\right) = \tfrac{1}{2}\cos\left(\varsigma_+(c)t - \tfrac{\pi}{4}\right) + \tfrac{1}{2}\cos\left(\varsigma_-(c)t + \tfrac{\pi}{4}\right) .$$

Since the saddle point of order two in $\hat{\theta}$ splits up into the two saddle points $\theta_\pm(c)$ of order one, we have two oscillating terms; each corresponding to one term in (3.1b).

Concerning the amplitude we expand $\tilde{\omega}''(\theta)$ and $c = \tilde{\omega}'(\theta)$ in $\hat{\theta}$, which implies $\tilde{\omega}''(\theta) = 2\tilde{\omega}'''(\hat{\theta})(\hat{c} - c) + \mathcal{O}\left((\hat{c} - c)^{2/3}\right)$. Thus, using the second representation of $a_{\hat{\theta}}$ from (3.5) yields

$$\frac{1}{\sqrt{\pi}}\mathbb{A}_{\hat{\theta}}(c)z^{-1/4}t^{-1/3} = \frac{1}{\sqrt{2\pi|\tilde{\omega}''(\theta)|}}\begin{pmatrix} 1 & \frac{1}{\omega_{\rm r}(\hat{\theta})} \\ \omega_{\rm r}(\hat{\theta}) & 1 \end{pmatrix} t^{-1/2} + \mathcal{O}(c - \hat{c}) .$$

Combining this with the oscillating terms we obtain two terms of (3.1b).

4 Asymptotic expansions - proofs

This section is dedicated to the proofs of the asymptotic expansions stated in the last section, namely to Theorem 3.3 and 3.1. Before starting the actual work, we outline the general strategy followed in Section 4.2 and 4.3 to derive the asymptotic expansions.

4.1 General strategy for the proofs

Consider an oscillatory integral of the form

$$g(t,c) = \int_{S^1} A(\theta)\mathrm{e}^{\mathrm{i}t\phi(\theta,c)}\,\mathrm{d}\theta \tag{4.1}$$

We aim to derive an asymptotic expansion which holds (locally) uniformly with respect to c, for instance for $c \in [c_* - \varepsilon, c_* + \varepsilon]$. The procedure splits into the following four steps.

Localization principle
As mentioned in Section 2.2, the asymptotic behavior of (4.1) for $t \to \infty$ is dominated by the wave numbers $\{\theta_1, ..., \theta_K\} = \boldsymbol{\Theta}(c_*)$ which fulfill the condition $\partial_\theta \phi(\theta_k, c_*) = 0$. This is due to fact to which [Ste93, VIII] refers as the first principle of oscillatory integrals, namely the localization principle. Here the underlying idea is that, if $|\partial_\theta \phi(\theta_k, c)|$ is uniformly bounded from below on $[\underline{\theta}, \overline{\theta}]$, we may apply partial integration to obtain

$$\left| \int_{\underline{\theta}}^{\overline{\theta}} A(\theta)\, \mathrm{e}^{\mathrm{i}t\phi(\theta,c)}\, \mathrm{d}\theta \right| = \left| \frac{A(\theta)}{\mathrm{i}t\partial_\theta\phi(\theta,c)} \right|_{\underline{\theta}}^{\overline{\theta}} - \int_{\underline{\theta}}^{\overline{\theta}} \partial_\theta \left(\frac{A(\theta)}{\mathrm{i}t\partial_\theta\phi(\theta,c)} \right) \mathrm{e}^{\mathrm{i}t\phi(\theta,c)}\, \mathrm{d}\theta \right| \le \frac{C}{t} \,.$$

If the boundary terms cancel due to periodicity or compact support we may iterate the argument to obtain bounds $\sim t^{-N}$. To utilize this fact for (4.1) we use a partition of unity $\{\psi_1, ..., \psi_K, 1 - \sum_k \psi_k\}$ on $\mathcal{S}^1$ with $\psi_k \in \mathcal{C}_0^k(\mathcal{U}_\delta(\theta_k))$ for some $\delta_k > 0$ which in general might depend on k. Then asymptotic behavior is dominated by terms of the form

$$I_k(t,c) = \int_{\theta_k - \delta}^{\theta_k + \delta} A(\theta)\psi_k(\theta)\mathrm{e}^{\mathrm{i}t\phi(\theta,c)}\, \mathrm{d}\theta \,. \tag{4.2}$$

For the localization error holds

$$|R_{loc}(t,c)| = \left| \int_{S^1} A(\theta) \Big[1 - \sum_k \psi_k(\theta)\Big] \mathrm{e}^{\mathrm{i}t\phi(\theta,c)}\, \mathrm{d}\theta \right| \le C\Big(\omega, \varepsilon, \min_k \delta_k\Big) t^{-N}$$

for some $N \in \mathbb{N}$. The bound $C(\omega, \varepsilon, \min_k \delta_k)$ depends on δ_k via $\|\psi_k\|_{W^{n,p}}$.

Local coordinate transformation
The next step consists in rewriting (4.2) using a suitable coordinate transform $u = U(\theta, c)$. In the simplest case with $|\partial_\theta^2 \phi(\theta_k, c)| = |\omega''(\theta_k)| \ne 0$, we may use the implicit function theorem to see that locally we can achieve $\phi(\theta, c) = \pm(u^2 + \phi(\theta_k, c))$. This applies in case of the inner regions, cf. Section 4.2. At the wave fronts the dispersion relation degenerates: For $\hat{\theta} \in \boldsymbol{\Theta}_{\mathrm{cr}}$ and $\hat{c} = c_{\mathrm{gr}}(\hat{\theta})$ holds $|\partial_\theta^2 \phi(\hat{\theta}, \hat{c})| = |\omega''(\hat{\theta})| = 0$ but $|\partial_\theta^2 \phi(\theta, c)| = |\omega''(\theta)| \ne 0$ for arbitrary c (in a sufficiently small neighborhood) if $\theta \ne \hat{\theta}$. In that case we may apply a suitable version of Weierstrass's Preparation theorem which is derived in appendix B. For instance, if $|\partial_\theta^3 \phi(\theta_k, c)| = |\omega'''(\theta_k)| \ne 0$ we obtain $\phi(\theta, c) = \frac{\sigma}{3}u^3 + a(c, \theta_k)u + b(c, \theta_k)$. The asymptotic behavior of (4.1) in this case is discussed in detail in Section 4.3. Since the preparation theorems are quite general the theory also applies for higher order of degeneracy.

In any case we may substitute $\theta = \Theta(u, c) := U^{-1}(u, c)$ in (4.2) such that

$$I_k(t,c) = \int_{-B}^{B} f(u,c)\, \mathrm{e}^{\mathrm{i}tp(u,c,\theta_k)}\, \mathrm{d}u$$

with $f(u) = A \circ \Theta(u,c) \cdot \psi_k \circ \Theta(u,c) \cdot \partial_u \Theta(u,c)$ and $p(\cdot, c, \theta_k)$ is a polynomial in a suitable normal form.

Actual decay rate
In this step the actual leading order term of the asymptotic expansion is derived by tracing back $I_k(t,c)$ to special functions. In the non-degenerated case we obtain integral of Fresnel-type, in the degenerated case we obtain Airy's function. In any case we basically obtain

$$I_k(t,c) = \mathcal{G}(t,c)\,t^{-1/n} + R_k(t,c)$$

with $n = 2$ or 3.

Estimation of the error terms
It remains to estimate the error term. We will prove that

$$|R_k(t,c)| \le C(\omega,\varepsilon)\,t^{-1}\,.$$

In general, this might not be optimal with respect to the decay in t. But we focus on the dependency which actually is more complicated. In fact, in the non-degenerated case C, as well as $\mathcal{G}$, becomes singular as $\varepsilon \to 0$. In this step we finally fix $\delta_k = \delta_k(\varepsilon)$.

Thus, combining all estimates we end up with

$$|g(t,c) - \mathcal{G}(t,c)\,t^{-1/n}| \le C(\omega,\varepsilon)\,t^{-1}\,.$$

4.2 Asymptotic expansion in the non-degenerate case

In this section we proof Theorem 3.1. We follow the strategy outlined in Section 4.1. The proof splits into two parts. First we apply standard method of stationary state, cf. for instance [Won89, Ste93] to derive the leading order term if the asymptotic expansion. Second we derive a uniform upper bound on the error term when the expansion becomes singular as $\varepsilon \to 0$.

We consider Green's function represented by (2.14). We choose a parametrization of $\mathcal{S}^1$ on $(2\pi, 2\pi]$, i.e. the components of $G(t,c)$ are given by

$$g(t,c) = \frac{1}{4\pi}\int_{-2\pi}^{2\pi} A(\theta)\,\mathrm{e}^{\mathrm{i}t\phi(\theta,c)}\,\mathrm{d}\theta \quad \text{with} \quad \phi(\theta,c) = c\theta - \tilde{\omega}(\theta), \quad c = \frac{j}{t}$$

and $A(\theta) = 1$, $\mathrm{e}^{\mathrm{i}\theta/2}/\omega_{\mathrm{r}}(\theta)$ or $\omega_{\mathrm{r}}(\theta)\mathrm{e}^{-\mathrm{i}\theta/2}$.

Assume $\varepsilon > 0$ and $c \in \omega'(\mathcal{S}^1) \setminus \mathcal{C}_{\mathrm{cr}}(\varepsilon)$. We implicitly exclude the degenerated case $\omega'(\mathcal{S}^1) \setminus \mathcal{C}_{\mathrm{cr}}(\varepsilon) = \emptyset$ where the statement becomes meaningless, i.e. we assume $\varepsilon < 2c_{\max}$. Assume $\Theta(c) = \{\theta_1, \ldots, \theta_K\}$.

Localization
For localization to the relevant wave numbers consider $\delta > 0$ such that

$$\forall\, k_1 \ne k_2 \;:\quad \mathcal{U}_\delta(\theta_{k_1}) \cap \mathcal{U}_\delta(\theta_{k_2}) = \emptyset \tag{4.3}$$

Without loss of generality we may assume $\mathcal{U}_\delta(\theta_k) \subset (-2\pi, 2\pi]$ for all $k = 1, \ldots, K$. Otherwise we simply may shift the parametrization of S^1. We choose a smooth partition of unity $\{\psi_1, \ldots, \psi_K, 1 - \sum \psi_k\}$ on $(-2\pi, 2\pi]$ with $\psi_k \in C_0^2(\mathcal{U}_\delta(\theta_k))$ and $\psi_k\big|_{\mathcal{U}_{\delta/2}(\theta_k)} = 1$. For later use we record that there exist $C_1, C_2 > 0$ such that

$$\|\psi_k\|_\infty = 1\,, \quad \|\psi_k'\|_\infty = \frac{C_1}{\delta}\,, \quad \|\psi_k''\|_\infty = \frac{C_2}{\delta^2}\,. \tag{4.4}$$

Now, the components of Green's function read as

$$g(t,c) = \frac{1}{4\pi} \sum_k \int_{\theta_k - \delta}^{\theta_k + \delta} \psi_k(\theta) A(\theta)\, \mathrm{e}^{\mathrm{i}t\phi(\theta,c)}\, \mathrm{d}\theta + R_{loc}(t,c) \tag{4.5}$$

with

$$R_{loc}(t,c) = \frac{1}{4\pi} \int_{-2\pi}^{2\pi} \Big[1 - \sum_k \psi_k(\theta)\Big] A(\theta)\, \mathrm{e}^{\mathrm{i}t\phi(\theta,c)}\, \mathrm{d}\theta\,. \tag{4.6}$$

Obviously, the localization error is $\mathcal{O}(t^{-N})$ for all $N \in \mathbb{N}$. But here we want to point out the dependency of the bound on ε. We postpone the discussion. First we derive the leading order asymptotic behavior. To do so we consider

$$I(t,\theta_k) = \int_{\theta_k}^{\theta_k + \delta} \psi_k(\theta) A(\theta) \mathrm{e}^{\mathrm{i}t\phi(\theta,c)}\,. \tag{4.7}$$

The discussion for $\int_{\theta_k - \delta}^{\theta_k}$ works analogous. Since from now on we will focus on on single integral of the this form, we omit the index k in ψ_k to simplify notation.

Local coordinate transform
We introduce a new variable u by

$$\phi(\theta,c) = -\operatorname{sign}\tilde{\omega}''(\theta_k) u^2 + \phi(\theta_k,c)$$

which defines the coordinate transform

$$U(\theta) = \sqrt{-\operatorname{sign}\tilde{\omega}''(\theta_k)\big(\phi(\theta,c) - \phi(\theta_k,c)\big)}\,.$$

The function $U(\theta)$ is smooth and monotone on $[\theta_k, \theta_{k+1})$ For later use we introduce

$$\begin{aligned} h(\theta,\theta_k) &= \frac{-2\operatorname{sign}\tilde{\omega}''(\theta_k)\big(\phi(\theta,c) - \phi(\theta_k,c)\big)}{(\theta-\theta_k)^2} - |\tilde{\omega}''(\theta_k)| \\ &= 2\operatorname{sign}\tilde{\omega}''(\theta_k) \sum_{n=1}^{\infty} \frac{\tilde{\omega}^{(n+2)}(\theta_k)}{(n+2)!}(\theta-\theta_k)^n \end{aligned} \tag{4.8}$$

such that we may write

$$U(\theta) = \tfrac{1}{\sqrt{2}}(\theta - \theta_k)\sqrt{|\tilde{\omega}''(\theta_k)| + h(\theta,\theta_k)}\,. \tag{4.9}$$

Here, the analyticity of $\tilde{\omega}$ carries over to h. Moreover, note that $h(\theta,\theta_k) = \mathcal{O}(\theta - \theta_k)$ as $\theta \to \theta_k$ and $U'(\theta_k) = \sqrt{\frac{|\tilde{\omega}''(\theta_k)|}{2}}$.

Using the implicit function theorem we invert the coordinate transform U, i.e. $\theta = \Theta(u) := U^{-1}(u)$ and substitute the integration variable in (4.7). Since ψ is compactly supported we may replace the upper bound $U(\theta_k+\delta)$ by ∞ such that

$$I(t,\theta_k) = \mathrm{e}^{\mathrm{i}t\,\mathrm{sign}\,\tilde{\omega}''(\theta_k)\phi(c,\theta_k)} \int_0^\infty f(u)\mathrm{e}^{\mathrm{i}tu^2}\,\mathrm{d}u \tag{4.10}$$

with $f(u) = A\circ\Theta(u)\cdot\psi\circ\Theta(u)\cdot\Theta'(u)$. Note that, via Θ and ψ, f depends on θ_k.

Actual decay rate

To derive the actual asymptotic expansion we use arguments based on partial integration. We basically follow the procedure presented in [Won89, II.3].

We define

$$K_0(u) = \mathrm{e}^{\mathrm{i}tu^2} \qquad \text{and} \qquad K_{n+1}(u) = -\int_u^{u+\infty\mathrm{e}^{\mathrm{i}\pi/4}} K_n(\tilde{u})\,\mathrm{d}\tilde{u}\,.$$

Thus, applying partial integration to (4.10) leads to

$$\begin{aligned}\mathrm{e}^{-\mathrm{i}t\,\mathrm{sign}\,\tilde{\omega}''(\theta_k)\phi(c,\theta_k)} I(t,\theta_k) &= f(u)K_1(u)\Big|_{u=0}^{\infty} - \int_0^\infty f'(u)K_1(u)\,\mathrm{d}u \\ &= -f(0)K_1(0) - \int_0^\infty f'(u)K_1(u)\,\mathrm{d}u\,.\end{aligned}$$

Using $f(0) = \psi(\theta_k)A(\theta_k)\Theta'(0) = A(\theta_k)\sqrt{\frac{|\tilde{\omega}''(\theta_k)|}{2}}$ and

$$K_1(0) = -\int_0^{\infty\mathrm{e}^{\mathrm{i}\pi/4}} \mathrm{e}^{\mathrm{i}tu^2}\,\mathrm{d}u = -\int_0^\infty \mathrm{e}^{-\frac{1}{2}u^2}\,\mathrm{d}u\cdot\frac{\mathrm{e}^{\mathrm{i}\frac{\pi}{4}}}{\sqrt{2t}} = -\frac{\sqrt{\pi}}{2}\,\mathrm{e}^{\mathrm{i}\frac{\pi}{4}}\,t^{-1/2}$$

we obtain the first order term of the asymptotic expansion,

$$I(t,\theta_k) = \mathrm{e}^{\mathrm{i}t\,\mathrm{sign}\,\tilde{\omega}''(\theta_k)\phi(c,\theta_k)} A(\theta_k)\sqrt{\frac{2\pi}{|\tilde{\omega}''(\theta_k)|}}\,\mathrm{e}^{\mathrm{i}\frac{\pi}{4}}\,t^{-1/2} + R(t,\theta_k)\,. \tag{4.11}$$

in which, after a second partial integration

$$\begin{aligned}\mathrm{e}^{-\mathrm{i}t\,\mathrm{sign}\,\tilde{\omega}''(\theta_k)\phi(c,\theta_k)} R(t,\theta_k) &= -\int_0^\infty f'(u)K_1(u)\,\mathrm{d}u \\ &= -f'(0)K_2(0) + \int_0^\infty f''(u)K_2(u)\,\mathrm{d}u\end{aligned}$$

holds for the error term. Thus we get

$$|R(t,\theta_k)| \le \frac{1}{2}\left(|f'(0)| + \int_0^\infty |f''(u)|\,\mathrm{d}u\right)t^{-1}\,. \tag{4.12}$$

Here the right hand side depends on ψ and via A and the coordinate transform Θ on the dispersion relation $\tilde{\omega}$ on $[\theta_k, \theta_k+\delta]$.

It remains to deter min how the right hand side depends on ε.Actually, this will take the larger part of the proof. Preliminary for that consider first the simplest example, namely the case of pure nearest neighbor interaction. In that case we have $\tilde{\omega}(\theta) = \frac{1}{2}\sin\frac{\theta}{2}$ and $\hat{c} = \pm 1$ which implies $|\tilde{\omega}''(\theta)| = \frac{1}{2}\sqrt{\hat{c}^2 - c^2} \sim \sqrt{\varepsilon}$. The next basic lemma states that this holds in general.

Lemma 4.1:
Consider the dispersion relation $\tilde{\omega}$ defined in (2.13) *and assume that the non-degeneracy condition* (2.10) *is satisfied. Then there exists $\delta_0 > 0$ and constants $\underline{C}(\tilde{\omega})$, $\overline{C}(\tilde{\omega}) > 0$ such that*

$$\forall \theta \in \bigcup_{\hat{\theta} \in \mathbf{\Theta}_{\mathrm{cr}}} \mathcal{U}_{\delta_0}(\hat{\theta}) : \quad \underline{C}(\tilde{\omega})|c - \hat{c}|^{1/2} \leq |\tilde{\omega}''(\theta)| \leq \overline{C}(\tilde{\omega})|c - \hat{c}|^{1/2} \tag{4.13}$$

where $c = \tilde{\omega}'(\theta)$.

Proof. According to the non-degeneracy condition we have $\tilde{\omega}'''(\hat{\theta}) \neq 0$ for all $\hat{\theta} \in \mathbf{\Theta}_{\mathrm{cr}}$. We define $\underline{\alpha} := \frac{1}{2}\min_{\hat{\theta} \in \mathbf{\Theta}_{\mathrm{cr}}} |\tilde{\omega}'''(\hat{\theta})|$ and $\overline{\alpha} := \max_{\theta \in S^1} |\tilde{\omega}'''(\theta)|$. Since $\mathbf{\Theta}_{\mathrm{cr}}$ is finite, there exists $\delta_0 > 0$ such that

$$\forall \theta \in \textstyle\bigcup_{\hat{\theta} \in \mathbf{\Theta}_{\mathrm{cr}}} \mathcal{U}_{\delta_0}(\hat{\theta}) : \quad \underline{\alpha} \leq |\tilde{\omega}'''(\theta)| \leq \overline{\alpha}.$$

As from now we assume $\theta \in \mathcal{U}_{\delta_0}(\hat{\theta})$ for some $\hat{\theta} \in \mathbf{\Theta}_{\mathrm{cr}}$. Due to $\tilde{\omega}''(\hat{\theta}) = 0$ it is straight forward that

$$\underline{\alpha}|\theta - \hat{\theta}| \leq |\tilde{\omega}''(\theta)| \leq \overline{\alpha}|\theta - \hat{\theta}| \,.$$

Now we consider $c - \hat{c} = \tilde{\omega}'(\theta) - \tilde{\omega}'(\hat{\theta})$ and find $|c - \hat{c}| \leq |\tilde{\omega}''(\theta)| \cdot |\theta - \hat{\theta}|$. Thus we obtain $|c - \hat{c}|^{1/2} \leq \frac{1}{\sqrt{\underline{\alpha}}}|\tilde{\omega}''(\theta)|$ which proves the first estimate of (4.13) with $\underline{C}(\tilde{\omega}) = \sqrt{\underline{\alpha}}$. To prove the second estimate note that, possibly after decreasing δ_0, $|c - \hat{c}| \geq \frac{1}{2}\underline{\alpha}|\theta - \hat{\theta}|^2$ again holds for all $\theta \in \mathcal{U}_{\delta_0}(\hat{\theta})$. Thus we get $|\tilde{\omega}''(\theta)| \leq \frac{\sqrt{2}\overline{\alpha}}{\sqrt{\underline{\alpha}}}|c - \hat{c}|^{1/2}$ which proves the second estimate in (4.13) with $\overline{C}(\tilde{\omega}) = \frac{\sqrt{2}\overline{\alpha}}{\sqrt{\underline{\alpha}}}$. □

Estimation of the error term $|R(t, \theta_k)|$
Now we are able to prove the second statement of Theorem 3.1. At first we determine a uniform bound on $|R(t, \theta_k)|$ for ε small where the asymptotic expansion becomes singular. To do so we first provide some formulas to express $f'(u)$ and $f''(u)$ in (4.12), respectively, in terms of the dispersion relation $\tilde{\omega}$. Obviously we have

$$f' = \psi A \Theta'' + (\psi_\theta A + \psi A_\theta)\,\Theta'^2 \tag{4.14}$$

$$f'' = \psi A \Theta''' + 3\,(\psi_\theta A + \psi A_\theta)\,\Theta'\Theta'' + (\psi_{\theta\theta} A + \psi A_{\theta\theta} + 2\psi_\theta A_\theta)\,\Theta'^3 \tag{4.15}$$

and, with $\theta = \Theta(u)$, for the coordinate transform holds

$$\Theta'(u) = \frac{1}{U'(\theta)}\,, \quad \Theta''(u) = -\frac{U''(\theta)}{U'(\theta)^3}\,, \quad \Theta'''(u) = \frac{3U''(\theta)^2 - U'(\theta)U'''(\theta)}{U'(\theta)^5}\,. \tag{4.16}$$

Finally we express the derivatives of U in terms of $\tilde{\omega}$ and h. According to (4.9) we find

$$
\begin{aligned}
U' &= \frac{2(|\tilde{\omega}''(\theta_k)|+h_k) + (\theta-\theta_k)h_k'}{2\sqrt{2}(|\tilde{\omega}''(\theta_k)|+h_k)^{1/2}} \\
U'' &= \frac{\big(4h_k' + 2h_k''(\theta-\theta_k)\big)(|\tilde{\omega}''(\theta_k)|+h_k) - {h_k'}^2(\theta-\theta_k)}{4\sqrt{2}(|\tilde{\omega}''(\theta_k)|+h_k)^{3/2}} \\
U''' &= \frac{\big(12h_k''+4h_k'''(\theta-\theta_k)\big)(|\tilde{\omega}''(\theta_k)|+h_k)^2}{8\sqrt{2}(|\tilde{\omega}''(\theta_k)|+h_k)^{5/2}} \\
&\quad + \frac{-\big(6{h_k'}^2+8h_k'h_k''(\theta-\theta_k)\big)(|\tilde{\omega}''(\theta_k)|+h_k) + 3{h_k'}^3(\theta-\theta_k)}{8\sqrt{2}(|\tilde{\omega}''(\theta_k)|+h_k)^{5/2}} \,.
\end{aligned}
\tag{4.17}
$$

Now we consider the first term on the right hand side of (4.12). Using (4.16), (4.17), (4.9) and $\psi'(\Theta(0)) = \psi'(\theta_k) = 0$ we find $\psi_\theta A{\Theta'}^2\big|_{u=0} = 0$, $\psi A_\theta {\Theta'}^2\big|_{u=0} = \frac{2A'(\theta_k)}{|\tilde{\omega}''(\theta_k)|}$ and $\psi A\Theta''\big|_{u=0} = \frac{2\,\mathrm{sign}\,\tilde{\omega}''(\theta_k)\,\tilde{\omega}'''(\theta_k)}{3|\tilde{\omega}''(\theta_k)|^2}$. Thus we have

$$
|f'(0)| \le \left| \frac{2A'(\theta_k)}{|\tilde{\omega}''(\theta_k)|} + \frac{2\,\mathrm{sign}\,\tilde{\omega}''(\theta_k)\,\tilde{\omega}'''(\theta_k)}{3|\tilde{\omega}''(\theta_k)|^2} \right| =: B_1(\theta_k)\,. \tag{4.18}
$$

Concerning the second term on the right hand side of (4.12), due to the compact support of ψ and $\int_0^{U(\theta_k+\delta)} A(\theta(u))\Theta'(u)\,\mathrm{d}u = \int_{\theta_k}^{\theta_k+\delta} A(\theta)\,\mathrm{d}\theta$, (4.15) leads to

$$
\begin{aligned}
\left|\int_0^\infty f''(u)\,\mathrm{d}u\right| \le& \|A\|_{W^{2,1}(\theta_k,\theta_k+\delta)} \Bigg(\left\| \frac{\Theta'''}{\Theta'} \right\|_{L^\infty(0,U(\theta_k+\delta))} \\
&+ 3\Big(1+\|\psi'\|_{L^\infty(\theta_k,\theta_k+\delta)}\Big)\|\Theta''\|_{L^\infty(0,U(\theta_k+\delta))} \\
&+ \Big(1+2\|\psi'\|_{L^\infty(\theta_k,\theta_k+\delta)}+\|\psi''\|_{L^\infty(\theta_k,\theta_k+\delta)}\Big)\|{\Theta'}^2\|_{L^\infty(0,U(\theta_k+\delta))} \Bigg)\,.
\end{aligned}
$$

Using (4.4) and (4.16) this implies

$$
|R(t,\theta_k)| \le \frac{1}{2}\left(B_1(\theta_k) + B_2(\theta_k,\delta) + \frac{B_3(\theta_k,\delta)}{\delta} + \frac{B_4(\theta_k,\delta)}{\delta^2} \right) t^{-1} \tag{4.19}
$$

with

$$
\begin{aligned}
B_2(\theta_k,\delta) &= \|A\|_{W^{2,1}(S^1)} \left\| \frac{3{U''}^2 - U'U'''}{{U'}^4} \right\|_{L^\infty(\theta_k,\theta_k+\delta)} \\
B_3(\theta_k,\delta) &= C\|A\|_{W^{2,1}(S^1)} \left\| \frac{U''}{{U'}^3} \right\|_{L^\infty(\theta_k,\theta_k+\delta)} \\
B_4(\theta_k,\delta) &= C\|A\|_{W^{2,1}(S^1)} \left\| \frac{1}{{U'}^2} \right\|_{L^\infty(\theta_k,\theta_k+\delta)}
\end{aligned}
\tag{4.20}
$$

and B_1 defined in (4.18).

Now recall the assumptions of the first part of the proof: $\varepsilon > 0$, $c \in \tilde{\omega}'(\mathcal{S}^1) \setminus \mathcal{C}_{\text{cr}}(\varepsilon)$, $\theta_k \in \{\theta_1, \ldots, \theta_K\} = \Theta(c)$ and $\delta > 0$ such that $\mathcal{U}_\delta(\theta_{k_1}) \cap \mathcal{U}_\delta(\theta_{k_2}) \neq \emptyset$ if $k_1 \neq k_2$ (cf. (4.3)) is fulfilled. We aim to prove that the bound on $R(t, \theta_k)$ is uniform with respect to $c \in \tilde{\omega}'(\mathcal{S}^1) \setminus \mathcal{C}_{\text{cr}}(\varepsilon)$. To do so we express the dependency on c in terms of θ_k. Since $\tilde{\omega}'$ is continuous and $\boldsymbol{\Theta}_{\text{cr}}$ finite, there exists $\delta_\varepsilon > 0$ such that $\tilde{\omega}'(\boldsymbol{\Theta}_{\text{cr}}(\delta_\varepsilon)) \subset \mathcal{C}_{\text{cr}}(\varepsilon)$. For later use we choose δ_ε maximal in the sense that there exists $\hat{\theta} \in \boldsymbol{\Theta}_{\text{cr}}$ such that $|\tilde{\omega}'(\hat{\theta}+\delta_\varepsilon) - \tilde{\omega}'(\hat{\theta})| = \varepsilon$ or $|\tilde{\omega}'(\hat{\theta}-\delta_\varepsilon) - \tilde{\omega}'(\hat{\theta})| = \varepsilon$. Thus, proving the assertion for $\theta_k \in \boldsymbol{\Theta}^c_{\text{cr}}(\delta_\varepsilon) = S^1 \setminus \boldsymbol{\Theta}_{\text{cr}}(\delta_\varepsilon)$ is sufficient since, due to $\tilde{\omega}'(\mathcal{S}^1) \setminus \mathcal{C}_{\text{cr}}(\varepsilon) \subset \tilde{\omega}'(\boldsymbol{\Theta}^c_{\text{cr}}(\delta_\varepsilon))$, this includes all demanded cases.

According to Lemma 4.1 there exists $\delta_0 > 0$ such that (4.13) holds. By eventually decreasing δ_0 we may ensure furthermore that $\boldsymbol{\Theta}_{\text{cr}}(\delta_0) = \overline{\bigcup_{\hat{\theta} \in \boldsymbol{\Theta}_{\text{cr}}} \mathcal{U}_\delta(\hat{\theta})}$ is a disjoint union. Note that this choice of δ_0 only depends on $\tilde{\omega}$. Now, to provide an upper bound on the right hand side of (4.19), we distinguish two cases.

First we consider $\theta_k \in \boldsymbol{\Theta}^c_{\text{cr}}(\delta_\varepsilon) \setminus \boldsymbol{\Theta}_{\text{cr}}(\delta_0)$. An upper bound on B_1 defined in (4.18) is provided by $B_1(\theta_k) \leq \max\left\{ B_1(\theta) | \theta \in \overline{S^1 \setminus \boldsymbol{\Theta}_{\text{cr}}(\delta_0)} \right\} := \bar{B}_1$ which only depends on $\tilde{\omega}$. To provide an upper bound on B_2, B_3 and B_4 we fix $\delta < \delta_0$, e.g. $\delta := \frac{\delta_0}{2}$. This is consistent with condition (4.3) since $\operatorname{dist}(\theta_k, \boldsymbol{\Theta}_{\text{cr}}) \geq \delta_0$ and the fact that there always exists $\hat{\theta} \in \boldsymbol{\Theta}_{\text{cr}}$ such that $\theta_k < \hat{\theta} < \theta_{k+1}$. We again have $B_n(\theta_k, \delta) \leq \max\left\{ B_n(\theta, \frac{\delta_0}{2}) | \theta \in \overline{S^1 \setminus \boldsymbol{\Theta}_{\text{cr}}(\delta_0)} \right\} := \bar{B}_n$. Thus, in case we proved the statement with a bound independently of ε,

$$\forall\, \theta_k \in \boldsymbol{\Theta}^c_{\text{cr}}(\delta_\varepsilon) \setminus \boldsymbol{\Theta}_{\text{cr}}(\delta_0): \quad |R(t, \theta_k)| \leq C(\tilde{\omega})\, t^{-1} \,. \tag{4.21}$$

In the second case, $\theta_k \in \boldsymbol{\Theta}^c_{\text{cr}}(\delta_\varepsilon) \cap \boldsymbol{\Theta}_{\text{cr}}(\delta_0)$, the singularity as $\varepsilon \to 0$ comes into play. The choice of δ_0 implies $\frac{1}{|\tilde{\omega}''(\theta_k)|} \leq \frac{1}{\underline{C}(\tilde{\omega})\varepsilon^{1/2}}$. Thus we obtain

$$B_1(\theta_k) \leq \frac{2|A'(\theta_k)|}{\underline{C}(\tilde{\omega})\,\varepsilon^{1/2}} + \frac{2|\tilde{\omega}'''(\theta_k)|}{3\,\underline{C}^2(\tilde{\omega})\,\varepsilon} \leq \frac{C(\tilde{\omega})}{\varepsilon} \,. \tag{4.22}$$

For B_2, B_3 and B_4 we have to provide lower bounds on the denominators on the right hand side of (4.20). To do so note that according to Lemma 4.1 holds $|\tilde{\omega}''(\theta_k)| \geq \underline{C}(\tilde{\omega})\varepsilon^{1/2}$. We introduce $v(\theta, \theta_k) := h(\theta, \theta_k) + \frac{1}{2}\partial_1 h(\theta, \theta_k)(\theta - \theta_k)$. Claiming that

$$\begin{aligned} &\forall \theta_k \in \boldsymbol{\Theta}^c_{\text{cr}}(\delta_\varepsilon) \cap \boldsymbol{\Theta}_{\text{cr}}(\delta_0)\ \forall \theta \in [\theta_k, \theta_k+\delta]: \\ &\qquad |h(\theta, \theta_k)| \leq \tfrac{1}{2}\,\underline{C}(\tilde{\omega})\,\varepsilon^{1/2} \quad \text{and} \quad |v(\theta, \theta_k)| \leq \tfrac{1}{2}\,\underline{C}(\tilde{\omega})\,\varepsilon^{1/2} \end{aligned} \tag{4.23}$$

we obtain

$$\begin{aligned} &|\tilde{\omega}''(\theta_k)| + h(\theta, \theta_k) \geq \tfrac{1}{2}\,\underline{C}(\tilde{\omega})\,\varepsilon^{1/2} \qquad \text{and} \\ &2\big(|\tilde{\omega}''(\theta_k)| + h(\theta, \theta_k)\big) + \partial_1 h(\theta, \theta_k)(\theta - \theta_k) \geq \underline{C}(\tilde{\omega})\,\varepsilon^{1/2} \,. \end{aligned} \tag{4.24}$$

To see that it is possible to choose $\delta > 0$ such that (4.23) holds consider $\tilde{h}(\Delta\theta, \theta_k) := h(\theta_k + \Delta\theta, \theta_k)$ on $[0, \delta_\varepsilon] \times \overline{\Theta^c_{\text{cr}}(\delta_\varepsilon) \cap \Theta_{\text{cr}}(\delta_0)}$. Since the domain is compact, $\tilde{h}$ continuous and $\tilde{h}(0, \theta_k) = 0$ holds for all θ_k, there exists $\delta_1 \in (0, \delta_\varepsilon]$ such that

$$\forall \Delta\theta \in [0, \delta_1] : \quad \max_{\theta_k} |\tilde{h}(\Delta\theta, \theta_k)| \leq \tfrac{1}{2} \underline{C}(\tilde{\omega})\, \varepsilon^{1/2} .$$

We may choose δ_1 maximal in the sense that either $\max_{\theta_k} |\tilde{h}(\Delta\theta, \theta_k)| \geq \frac{1}{2} \underline{C}(\tilde{\omega})\, \varepsilon^{1/2}$ for $\Delta\theta \in [\delta_1, \tilde{\delta}_1)$ with some $\tilde{\delta}_1 > 0$ or $\delta_1 = \delta_\varepsilon$. Repeating the arguments for $\tilde{v}(\Delta\theta, \theta_k) := v(\theta, \theta_k)$ leads to $\delta_2 > 0$. Choosing $\delta := \min\{\delta_1, \delta_2\}$ provides (4.23). Now, using (4.17) and $\varepsilon \leq 2 \max_{\theta \in \mathcal{S}^1} |\tilde{\omega}'(\theta)|$ we obtain

$$\begin{aligned} &\left\| \frac{3U''(\theta)^2 - U'(\theta)U'''(\theta)}{U'(\theta)^4} \right\|_{L^\infty(\theta_k, \theta_k+\delta)} \leq \frac{C(\tilde{\omega})}{\varepsilon^{5/2}} , \\ &\left\| \frac{U''(\theta)}{U'(\theta)^3} \right\|_{L^\infty(\theta_k, \theta_k+\delta)} \leq \frac{C(\tilde{\omega})}{\varepsilon^{3/2}} , \quad \text{and} \quad \left\| \frac{1}{U'(\theta)} \right\|^2_{L^\infty(\theta_k, \theta_k+\delta)} \leq \frac{C(\tilde{\omega})}{\varepsilon^{1/2}} \end{aligned} \tag{4.25}$$

with constants $C(\tilde{\omega}) > 0$ depending only on $\tilde{\omega}$.

It remains to express δ in terms of ε. To do so we distinguish three cases, namely $\delta = \delta_1 < \delta_\varepsilon$, $\delta = \delta_2 < \delta_\varepsilon$ and $\delta = \delta_\varepsilon$. In the first case we consider $\bar{\theta}_k := \operatorname{argmax} |\tilde{h}(\delta, \theta_k)|$. In view of (4.9) an the fact that $\tilde{\omega}$ is real analytic $\tilde{h}$ in fact is well defined and smooth on a compact domain which is a superset of $[0, \delta_\varepsilon] \times \overline{\Theta^c_{\text{cr}}(\delta_\varepsilon) \cap \Theta_{\text{cr}}(\delta_0)}$ and independent of ε, for instance $[0, \pi] \times [-\pi, \pi]$. On this set there exists a global Lipschitz constant $L_{\tilde{h}} = L_{\tilde{h}}(\tilde{\omega})$ only depending on $\tilde{\omega}$. We obtain $\frac{1}{2} \underline{C}(\tilde{\omega})\, \varepsilon^{1/2} = |\tilde{h}(\delta, \bar{\theta}_k)| \leq L_{\tilde{h}}(\tilde{\omega})\, \delta$, i.e. $\frac{1}{\delta} \leq \frac{2L_{\tilde{h}}(\tilde{\omega})}{\underline{C}(\tilde{\omega})\, \varepsilon^{1/2}} = \frac{C(\tilde{\omega})}{\varepsilon^{1/2}}$. In the second case we again repeat the arguments for $\delta = \delta_2$ and $\tilde{v}(\Delta\theta, \theta_k)$. For the third case recall that there exists $\hat{\theta} \in \Theta_{\text{cr}}$ such that $|\tilde{\omega}(\hat{\theta} \pm \delta_\varepsilon) - \tilde{\omega}(\hat{\theta})| = \varepsilon$. According to Lemma 4.1 we have $\varepsilon = |\tilde{\omega}(\hat{\theta} \pm \delta_\varepsilon) - \tilde{\omega}(\hat{\theta})| \leq \max_{\theta \in [\hat{\theta}, \hat{\theta}+\delta_\varepsilon]} |\tilde{\omega}''(\theta)|\, \delta_\varepsilon \leq \overline{C}(\tilde{\omega})\, \varepsilon^{1/2}\, \delta_\varepsilon$ which leads to the same order of decay as in the first two cases, i.e. we have

$$\frac{1}{\delta} \leq \frac{C(\tilde{\omega})}{\varepsilon^{1/2}} . \tag{4.26}$$

To summarize, the terms on the right hand side of (4.19) are bounded as follows, $B_1(\theta_k) \leq \frac{C(\tilde{\omega})}{\varepsilon^{1/2}}$, $B_2(\theta_k) \leq \frac{C(\tilde{\omega})}{\varepsilon^{5/2}}$, $\frac{B_3(\theta_k)}{\delta} \leq \frac{C(\tilde{\omega})}{\varepsilon^2}$ and $\frac{B_4(\theta_k)}{\delta^2} \leq \frac{C(\tilde{\omega})}{\varepsilon^{3/2}}$. Thus we proved

$$\forall\, \theta_k \in \Theta^c_{\text{cr}}(\delta_\varepsilon) \cap \Theta_{\text{cr}}(\delta_0) : \quad |R(t, \theta_k)| \leq C(\tilde{\omega}) \varepsilon^{-5/2}\, t^{-1} \tag{4.27}$$

Combining all bounds we obtain

$$|R(t, \theta_k)| \leq \frac{1}{2} \left(C_1(\tilde{\omega}) + \frac{C_2(\tilde{\omega})}{\varepsilon^{5/2}} \right) t^{-1} . \tag{4.28}$$

Estimation of the error term $|R_{loc}(t, c)|$
Finally we consider the localization error given in (4.6). Using the notation $\theta_k <$

$\hat{\theta}_k < \theta_{k+1}$ we see that the right hand side of (4.6) decomposes into terms of the form

$$\int_{\theta_k+\frac{\delta}{2}}^{\hat{\theta}_k} \left[1-\psi_k(\theta)\right] A(\theta)\, \mathrm{e}^{it\phi(\theta,c)}\, \mathrm{d}\theta$$

Note that $|\phi'(\theta,c)|$ strictly monotone on $[\theta_k,\hat{\theta}_k]$. Using Lemma 4.1 and (4.26) we find $|\phi'(\theta,c)| \geq \left|\phi'\left(\theta_k+\frac{\delta}{2},c\right)\right| \geq \min_{\theta\in\mathcal{S}^1\setminus\boldsymbol{\Theta}_{\mathrm{cr}}(\frac{\delta_\varepsilon}{2})} |\tilde{\omega}''(\theta)|\,\frac{\delta}{2} \geq \min\left\{C_1(\tilde{\omega}), C_2(\tilde{\omega})\,\varepsilon\right\}$. Thus, applying partial integration we find

$$\left|\int_{\theta_k+\frac{\delta}{2}}^{\hat{\theta}_k} \left[1-\psi_k(\theta)\right] A(\theta)\, \mathrm{e}^{it\phi(\theta,c)}\, \mathrm{d}\theta\right| \leq \left(C_1(\tilde{\omega}) + \frac{C_2(\tilde{\omega})}{\varepsilon}\right) t^{-1}.$$

This decay rate is already covered by (4.28) which completes the proof.

4.3 Asymptotic expansion at the wave fronts

This section is dedicated to the proof of Theorem 3.3. Arguments presented in [Hör90, Won89, BH86] are underlying the following considerations.

Consider again Green's function represented by (2.14). We choose a parametrization of $\mathcal{S}^1$ on $(-2\pi, 2\pi]$, i.e. the components of $G(t,c)$ are given by

$$g(t,c) = \frac{1}{4\pi}\int_{-2\pi}^{2\pi} A(\theta)\mathrm{e}^{\mathrm{i}t\phi(\theta,c)}\,\mathrm{d}\theta \quad \text{with} \quad \phi(\theta,c) = c\theta - \tilde{\omega}(\theta), \quad c = \frac{j}{t}$$

and $A(\theta) = 1$, $\frac{\mathrm{e}^{\mathrm{i}\frac{\theta}{2}}}{\omega_{\mathrm{r}}(\theta)}$ or $\omega_{\mathrm{r}}(\theta)\mathrm{e}^{-\mathrm{i}\frac{\theta}{2}}$.

Localization

We aim to determine the asymptotic behavior of $g(t,c)$ as $t \to \infty$ for group velocities c near caustics, i.e. $|c-\hat{c}| = \varepsilon$ with $\hat{c} = \tilde{\omega}'(\hat{\theta})$, $\hat{\theta} \in \boldsymbol{\Theta}_{\mathrm{cr}}$. Here, in general, the asymptotic behavior splits up into two components. One contribution $\sim t^{-1/3}$ is made by a neighborhood of the actual critical wave number $\hat{\theta}$ and one $\sim t^{-1/2}$ by further wave numbers $\theta \in \boldsymbol{\Theta}(c)$ which are bounded away from that. The latter vanishes if we are close to the sonic wave speeds $\pm c_{\max} = \pm\max_{\theta\in\boldsymbol{\Theta}_{\mathrm{cr}}} \tilde{\omega}'(\hat{\theta})$. To formalize this define

$$\delta_0 := \tfrac{1}{4} \min_{\hat{\theta}_1\neq\hat{\theta}_2\in\boldsymbol{\Theta}_{\mathrm{cr}}} |\hat{\theta}_1 - \hat{\theta}_2| \qquad \text{and} \qquad \varepsilon_0 := \min_{\hat{\theta}\in\boldsymbol{\Theta}_{\mathrm{cr}}} \left\{\tfrac{1}{2}\left|\tilde{\omega}'(\hat{\theta}) - \tilde{\omega}'(\hat{\theta}\pm\delta_0)\right|\right\}.$$

Note that both are uniquely determined by $\tilde{\omega}$, i.e. bounds depending beside $\tilde{\omega}$ on δ_0 or ε_0 are according to our convention denoted by $C(\tilde{\omega})$.

Now consider $c \in \mathcal{C}_{\mathrm{cr}}(\varepsilon_0)$ and $\{\hat{\theta}_1,\ldots,\hat{\theta}_K\} = \boldsymbol{\Theta}_{\mathrm{cr}}(c,\varepsilon_0)$. To simplify notation we assume $\mathcal{U}_{2\delta_0}(\hat{\theta}) \subset (-2\pi, 2\pi]$ for all k. For localization we use $\psi_k \in C_0^\infty(\mathcal{U}_{2\delta_0}(\hat{\theta}_k))$ with $\psi|_{[\hat{\theta}_k-\delta_0,\hat{\theta}_k+\delta_0]} \equiv 1$ and obtain

$$g(t,c) = \frac{1}{4\pi}\sum_k \int_{\hat{\theta}_k-\delta_0}^{\hat{\theta}_k+\delta_0} A(\theta)\,\mathrm{e}^{\mathrm{i}t\phi(\theta,c)}\,\mathrm{d}\theta + R_0(t,c) + R_{loc}(t,c) \tag{4.29}$$

with

$$R_0(t,c) = \frac{1}{4\pi}\sum_k \left(\int_{\hat\theta_k-2\delta_0}^{\hat\theta_k-\delta_0} + \int_{\hat\theta_k+\delta_0}^{\hat\theta_k+2\delta_0}\right) \psi_k(\theta)A(\theta)\,\mathrm{e}^{\mathrm{i}t\phi(\theta,c)}\,\mathrm{d}\theta$$

$$R_{loc}(t,c) = \frac{1}{4\pi}\left(\int_{-\pi}^{\hat\theta_1-\delta_0} + \sum_{k=1}^{K-1}\int_{\hat\theta_k+\delta_0}^{\hat\theta_{k+1}-\delta_0} + \int_{\hat\theta_K+\delta_0}^{\pi}\right)\Big[1-\sum_k \psi_k(\theta)\Big]A(\theta)\,\mathrm{e}^{\mathrm{i}t\phi(\theta,c)}\,\mathrm{d}\theta\,.$$

First we treat the error term R_0. In view of our choice of δ_0, $|\hat c - \tilde\omega'(\theta)|$ is monotonically increasing on $[\hat\theta_k, \hat\theta_k{+}2\delta_0]$. Utilizing the definition of ε_0 we find that $|\partial_\theta\phi(\theta,c)| \geq -|\hat c - c| + |\hat c - \tilde\omega'(\theta)| \geq \varepsilon_0$ for $\theta \in [\hat\theta_k{+}\delta_0, \hat\theta_k{+}2\delta_0]$. Since the same estimate holds on $[\hat\theta_k{-}2\delta_0, \hat\theta_k]$, integration by parts leads to

$$|R_0(t,c)| \leq C(\tilde\omega)\,t^{-1}\,.$$

Introducing $R_0(t,c)$ would actually not have been necessary to determine the asymptotic expansion since the upcoming steps also do the job if it is merged with the remaining integral on the right hand side of (4.29), cf. [Hör90]. But by doing so we may treat the remaining integral in an analytic setting which makes it more easy to determine the leading order error terms explicitly. Here the pay-off are error terms $\sim t^{-1}$ at the boundary $\hat\theta{\pm}\delta_0$. But this is what we get anyway from R_{loc}. We also could have used a simple cut-off, but we introduced ψ_k to highlight the origin of the contribution.

Concerning the localization error R_{loc} we may distinguish two cases. First, assume there exists $\theta \in \boldsymbol{\Theta}(c) \setminus \{\theta \,|\, \mathrm{dist}(\boldsymbol{\Theta}_{\mathrm{cr}}(c,\varepsilon_0),\theta) \leq \delta_0\}$. Since we are again uniformly bounded away from other critical wave numbers, we may use the theory presented in Section 4.2 to prove

$$\big|R_{loc}(t,c) - \mathcal{G}_{\mathrm{non}}(t,c)t^{-1/2}\big| \leq C(\tilde\omega)\,t^{-1}\,.$$

Here $\mathcal{G}_{\mathrm{non}}(t,c)$ is defined in (3.1b), where in this case we only sum over $\boldsymbol{\Theta}(c) \setminus \boldsymbol{\Theta}_{\mathrm{cr}}(\delta_0)$. In the second case, if $\theta \in \boldsymbol{\Theta}(c) \setminus \{\theta \,|\, \mathrm{dist}(\boldsymbol{\Theta}_{\mathrm{cr}}(c,\varepsilon_0),\theta) \leq \delta_0\} = \emptyset$, we are obviously near the sonic wave speed. We again have $|\partial_\theta\phi| \geq \varepsilon_0$ but this time iterative application of partial integration leads to

$$|R_{loc}(t,c)| \leq C_N(\tilde\omega)\,t^{-N} \qquad \text{for all } N \in \mathbb{N}$$

since the boundary terms cancel due to periodicity and the compact support of ψ_k, respectively.

Local coordinate transform

Since from now on we will consider only one single integral $\int_{\hat\theta_k-\delta_0}^{\hat\theta_k+\delta_0} A(\theta)\,\mathrm{e}^{\mathrm{i}t\phi(\theta,c)}\,\mathrm{d}\theta$ we skip the indices k and $\hat\theta$. Using the non-degeneracy condition $\omega'''(\hat\theta) \neq 0$ we can use the following proposition summarizing all the necessary results concerning the local coordinate transform, cf. [BH86, Won89] for a similar procedure.

Proposition 4.2 (Coordinate change for wave front):
Assume that $\theta \mapsto \omega(\theta)$ is real analytic near $\hat{\theta}$ and that

$$\hat{c} := \omega'(\hat{\theta}), \quad \omega''(\hat{\theta}) = 0, \quad \text{and} \quad \omega'''(\hat{\theta}) \neq 0.$$

Then, for the function $\phi(\theta, \hat{c}) = c\theta - \omega(\theta)$ there exists a unique local, analytical coordinate change $u = U(\theta, c)$ near $(\hat{\theta}, \hat{c})$ with inverse $\theta = \Theta(u, c)$ satisfying $U(\hat{\theta}, \hat{c}) = 0$, $\partial_\theta U(\hat{\theta}, \hat{c}) > 0$ and

$$\begin{aligned} &\phi(\theta, c) = \phi(\Theta(u, c), c) = -\frac{\sigma}{3}u^3 - a(c)u + b(c) \quad \text{with} \\ &\sigma := \operatorname{sign}\big(\omega'''(\hat{\theta})\big) \in \{-1, 1\}, \quad a(\hat{c}) = 0, \quad \sigma a'(\hat{c}) > 0, \quad b(\hat{c}) = \phi(\hat{\theta}, \hat{c}). \end{aligned}$$

The functions a and b can explicitly be constructed as follows. For $0 < \sigma(c - \hat{c}) \ll 1$ the equation $0 = \partial_\theta \phi(\theta, c) = c - \omega'(\theta)$ has exactly two solutions near $\hat{\theta}$, namely $\theta_-(c) < \theta_+(c)$. Now, for $0 < \sigma(c - \hat{c}) \ll 1$ the functions a and b are given by

$$a(c) = \sigma\Big(\frac{3\sigma}{4}\big[\phi(\theta_+(c), c) - \phi(\theta_-(c), c)\big]\Big)^{2/3}, \qquad b(c) = \frac{1}{2}\big[\phi(\theta_+(c), c) + \phi(\theta_-(c), c)\big].$$

In particular, if ϕ is odd in θ (i.e. $\phi(-\theta, c) = -\phi(\theta, c)$), then $b \equiv 0$.

Remark 4.3:
Before we give the proof of this result, we show how the result looks like for the FPU chain with $\omega(\theta) = 2\sin(\theta/2)$ with $\hat{\theta} = 0$, where $\hat{c} = 1$ and $\omega'''(0) = -1/4$. We find $\theta_\pm(c) = \pm 2\arccos(c)$ for $0 < c < \hat{c} = 1$. By symmetry we have $b \equiv 0$ and find $a(c) = (3[\sqrt{1-c^2} - c\arccos c])^{2/3}$. This function can be extended analytically into a neighborhood of $\hat{c}$ as follows. With $\alpha = \arccos c$ we have

$$a(c) = (3[\sin\alpha - \alpha\cos\alpha])^{2/3} = \alpha^2 g(\alpha^2)^{2/3},$$

where g is analytic around 0 with $g(0) = 1$ by using that $3[\sin\alpha - \alpha\cos\alpha] = \alpha^3(1 + \sum_{k=1}^\infty c_k \alpha^{2k})$. Now defining T as the inverse of $C(r) = \sum_{k=0}^\infty (-r)^k/((2k)!)$ (i.e. $C(r^2) = \cos r$ and $C(-r^2) = \cosh r$) we find $a(c) = T(c)g(T(c))^{2/3}$, which is real analytic for $c \in (-1, c_*)$ with $c_* \approx 44.7$, see Figure 4.1.

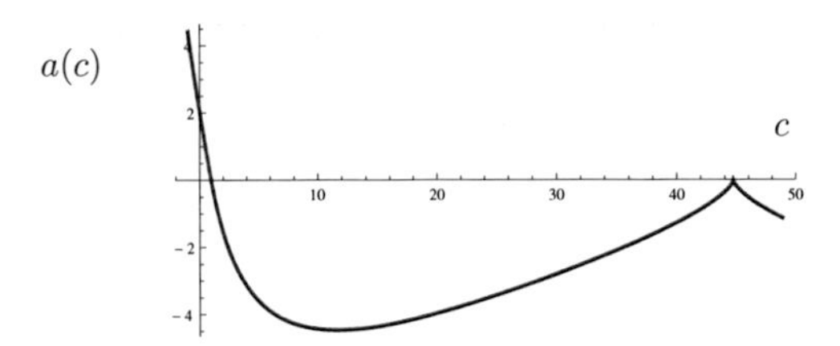

Figure 4.1: Function $a(c)$ for FPU dispersion relation $\omega(\theta) = 2\sin(\theta/2)$ at $\hat{\theta} = 0$.

Proof. Without loss of generality we carry out the details of the proof for the case that $\tilde{\omega}'$ takes a local maximum at $\hat{\theta}$, i.e. $\tilde{\omega}'''(\hat{\theta}) < 0$. We introduce $\varepsilon := \hat{c} - c$. Due to $c = \tilde{\omega}'(\theta)$ we have $\varepsilon \geq 0$ in the considered neighborhood of $\hat{\theta}$. For

$$\hat{\phi}(\theta, \varepsilon) := \phi(\theta, \hat{c}-\varepsilon) - \phi(\hat{\theta}, \hat{c}) = -\theta\varepsilon - \tilde{\omega}(\theta) + \tilde{\omega}(\hat{\theta}) + \tilde{\omega}'(\hat{\theta})(\theta - \hat{\theta})$$

holds

$$\hat{\phi}(\hat{\theta}, 0) = \partial_\theta \hat{\phi}(\hat{\theta}, 0) = \partial_\theta^2 \hat{\phi}(\hat{\theta}, 0) = 0 \qquad \text{and} \qquad \partial_\theta^3 \hat{\phi}(\hat{\theta}, 0) > 0 \, .$$

Thus we are able to apply the suitable version of Weierstrass preparation theorem yielding a normal form for $\hat{\phi}$. According to Theorem B.5, in a neighborhood of the critical point $\hat{\theta}$, there exists a coordinate transform

$$u = U(\theta, \varepsilon) \text{ or } \theta = \Theta(u, \varepsilon): \quad \hat{\phi}(\Theta(u, \varepsilon), \varepsilon) = \Phi(u, \varepsilon) := \tfrac{1}{3}u^3 - a(\varepsilon)u + b(\varepsilon) \, . \tag{4.30}$$

The transform $U(\theta, \varepsilon)$ as well as $a(\varepsilon)$ and $b(\varepsilon)$ are real analytic. Furthermore we have $a(0) = b(0) = 0$.

In case $\hat{\phi}$ is symmetric in $\hat{\theta}$, we have $b(\varepsilon) \equiv 0$ such that the oscillatory contribution depends on $\hat{\theta}$ only.

To make a and b explicit we use that, for $\theta \neq \hat{\theta}$, the saddle point of order 2 splits up into two saddle points of order 1. Thus, for $\varepsilon > 0$ there exist $\theta_\pm(\varepsilon)$ such that $\hat{c} - \tilde{\omega}'(\hat{\theta}_\pm(\varepsilon)) = \varepsilon$, which means $\partial_\theta \hat{\phi}(\theta_\pm(\varepsilon), \varepsilon) = 0$ for all ε. We now set $u_\pm(\varepsilon) := U(\theta_\pm(\varepsilon), \varepsilon)$ and observe

$$\left(u_\pm(\varepsilon)\right)^2 - a(\varepsilon) = \partial_u \Phi(u_\pm(\varepsilon), \varepsilon) = \partial_\theta \hat{\phi}(\theta_\pm(\varepsilon), \varepsilon)\, \partial_u \Theta(u_\pm(\varepsilon), \varepsilon) = 0.$$

Hence, we conclude $u = \pm\sqrt{a(\varepsilon)}$, and using the definition of Φ in (4.30) we obtain

$$a(\varepsilon) = \left(\frac{3}{4}\Big[\hat{\phi}(\theta_+(\varepsilon), \varepsilon) - \hat{\phi}(\theta_-(\varepsilon), \varepsilon)\Big]\right)^{2/3}, \quad b(\varepsilon) = \frac{1}{2}\Big[\hat{\phi}(\theta_+(\varepsilon), \varepsilon) + \hat{\phi}(\theta_-(\varepsilon), \varepsilon)\Big] \, .$$

Expanding $\theta_\pm(\varepsilon)$ in powers of $\varepsilon^{1/2}$ gives

$$\theta_\pm(\varepsilon) = \hat{\theta} \pm \sqrt{\frac{2}{-\tilde{\omega}'''(\hat{\theta})}}\, \varepsilon^{1/2} + \frac{\tilde{\omega}^{(4)}(\hat{\theta})}{3\tilde{\omega}'''(\hat{\theta})^2}\, \varepsilon + \mathcal{O}(\varepsilon^{3/2}). \tag{4.31}$$

Inserting this into $\hat{\phi}(\theta_\pm(\varepsilon), \varepsilon)$ leads to the expansions

$$a(\varepsilon) = -\sqrt[3]{\tfrac{2}{\tilde{\omega}'''(\hat{\theta})}}\, \varepsilon + \mathcal{O}(\varepsilon^2) \, , \qquad b(\varepsilon) = 2\hat{\theta}\, \varepsilon + \mathcal{O}(\varepsilon^2) \, .$$

We derived these identities for $\varepsilon \in [0, \varepsilon_0]$. But since a and b are real analytic this actually holds at least for $\varepsilon \in [-\varepsilon_0, \varepsilon_0]$.

Concerning the symmetric case where we have $b \equiv 0$ note that $\hat{\phi}(\hat{\theta}-\Delta\theta, c) = -\hat{\phi}(\hat{\theta}+\Delta\theta, c)$ implies $2\hat{\theta}c = \tilde{\omega}(\hat{\theta}+\Delta\theta) + \tilde{\omega}(\hat{\theta}-\Delta\theta)$. Keeping in mind that $c = j/t$ and that fact that we actually consider $\mathrm{e}^{it\phi}$ we find that the last identity needs to be valid $\bmod 2\pi/t$. I.e. since c is arbitrary, we have $b \equiv 0$ if $\hat{\theta} = 0 \bmod \pi$. □

Actual decay rate
Obviously, $\mathcal{U}_{\delta_0}(\hat{\theta}) \times [-\varepsilon_0, \varepsilon_0]$ is included in the neighborhood where the statement holds. Introducing the new variable u we may rewrite the integral on the right hand side of (4.29) as follows,

$$\begin{aligned} I(t,\varepsilon) &= \int_{\hat{\theta}-\delta_0}^{\hat{\theta}+\delta_0} A(\theta)\, \mathrm{e}^{\mathrm{i}t\phi(\theta,\hat{c}-\varepsilon)}\, \mathrm{d}\theta \\ &= \mathrm{e}^{\mathrm{i}t[b(\varepsilon)-\phi(\hat{\theta},\hat{c})]} \int_{U(\hat{\theta}-\delta_0,\varepsilon)}^{U(\hat{\theta}+\delta_0,\varepsilon)} A\big(\Theta(u,\varepsilon)\big)\, \partial_u\Theta(u,\varepsilon)\, \mathrm{e}^{\mathrm{i}t(u^3/3-a(\varepsilon)u)}\, \mathrm{d}u\,. \end{aligned} \tag{4.32}$$

To calculate the actual decay rate we again apply a version of Weierstrass division theorem. Regarding $f(u,\varepsilon) := A\big(\Theta(u,\varepsilon)\big)\, \partial_u\Theta(u,\varepsilon)$, Theorem B.2 yields

$$f(u,\varepsilon) = q(u,\varepsilon)\,\big[u^2 - a(\varepsilon)\big] + \beta(\varepsilon)u + \alpha(\varepsilon) \tag{4.33}$$

with real analytic functions q, β and α. Thus, (4.32) reads as

$$I(t,\varepsilon) = \mathrm{e}^{\mathrm{i}t[b(\varepsilon)-\phi(\hat{\theta},\hat{c})]} \int_{U(\hat{\theta}-\delta_0,\varepsilon)}^{U(\hat{\theta}+\delta_0,\varepsilon)} \big[\beta(\varepsilon)u + \alpha(\varepsilon)\big]\mathrm{e}^{\mathrm{i}t(u^3/3-a(\varepsilon)u)}\, \mathrm{d}u + R_1(t,\varepsilon) \tag{4.34}$$

with

$$\begin{aligned} R_1(t,\varepsilon) &= \mathrm{e}^{\mathrm{i}t[b(\varepsilon)-\phi(\hat{\theta},\hat{c})]} \int_{U(\hat{\theta}-\delta_0,\varepsilon)}^{U(\hat{\theta}+\delta_0,\varepsilon)} q(u,\varepsilon)\,\big[u^2 - a(\varepsilon)\big]\, \mathrm{e}^{\mathrm{i}t(u^3/3-a(\varepsilon)u)}\, \mathrm{d}u \\ &= \frac{-\mathrm{i}}{t}\, \mathrm{e}^{\mathrm{i}t[b(\varepsilon)-\phi(\hat{\theta},\hat{c})]} \Bigg(q(u,\varepsilon)\, \mathrm{e}^{\mathrm{i}t(u^3/3-a(\varepsilon)u)}\Big|_{U(\hat{\theta}-\delta_0,\varepsilon)}^{U(\hat{\theta}+\delta_0,\varepsilon)} \\ &\qquad - \int_{U(\hat{\theta}-\delta_0,\varepsilon)}^{U(\hat{\theta}+\delta_0,\varepsilon)} \partial_u q(u,\varepsilon)\, \mathrm{e}^{\mathrm{i}t(u^3/3-a(\varepsilon)u)}\, \mathrm{d}u \Bigg) \end{aligned} \tag{4.35}$$

The first two terms on the right hand side of (4.34) bear the leading order asymptotic behavior. We have

$$\int_{U(\hat{\theta}-\delta_0,\varepsilon)}^{U(\hat{\theta}+\delta_0,\varepsilon)} \mathrm{e}^{\mathrm{i}t(u^3/3-a(\varepsilon)u)}\, \mathrm{d}u = 2\pi\, \mathrm{Ai}\big(-a(\varepsilon)t^{2/3}\big)\, t^{-1/3} + R_2(t,\varepsilon) \quad \text{and}$$

$$\int_{U(\hat{\theta}-\delta_0,\varepsilon)}^{U(\hat{\theta}+\delta_0,\varepsilon)} u\, \mathrm{e}^{\mathrm{i}t(u^3/3-a(\varepsilon)u)}\, \mathrm{d}u = 2\pi\mathrm{i}\, \mathrm{Ai}'\big(-a(\varepsilon)t^{2/3}\big)\, t^{-2/3} + R_3(t,\varepsilon),$$

$$\text{where } R_j(t,\varepsilon) = \int_{\mathbb{R}\setminus[U(\hat{\theta}-\delta_0,\varepsilon),U(\hat{\theta}+\delta_0,\varepsilon)]} u^{j-2}\, \mathrm{e}^{\mathrm{i}t(u^3/3-a(\varepsilon)u)}\mathrm{d}u \quad \text{for } j = 2,3\,.$$

Here $\mathrm{Ai}(\cdot)$ denotes Airy's function $\mathrm{Ai}(z) = \frac{1}{2\pi}\int_{\mathbb{R}} \mathrm{e}^{i(u^3/3+zu)}\mathrm{d}u$.

Before estimating the error terms we want to make the last two functions on the right hand side of (4.33) explicit to determine the leading order terms. To do so note that

$$f(\pm\sqrt{a(\varepsilon)},\varepsilon) = \pm\beta(\varepsilon)\sqrt{a(\varepsilon)} + \alpha(\varepsilon)$$

which implies

$$\alpha(\varepsilon) = \frac{1}{2}\left[f(\sqrt{a(\varepsilon)},\varepsilon) + f(-\sqrt{a(\varepsilon)},\varepsilon)\right], \quad \beta(\varepsilon) = \frac{1}{2\sqrt{a(\varepsilon)}}\left[f(\sqrt{a(\varepsilon)},\varepsilon) - f(-\sqrt{a(\varepsilon)},\varepsilon)\right]$$

and $q(u,\varepsilon) = \frac{f(u,\varepsilon)-\beta(\varepsilon)u-\alpha(\varepsilon)}{u^2-a(\varepsilon)}$. The first factor of $f(\pm\sqrt{a(\varepsilon)},\varepsilon)$ we have at hand explicitly. We either have $A \equiv 1$ or, in view of (4.31), $A\big(\Theta(\pm\sqrt{a(\varepsilon)},\varepsilon)\big) = A(\hat{\theta}+\Delta\theta_\pm) = A(\hat{\theta}) \pm \tilde{A}_1\varepsilon^{1/2} + \tilde{A}_2\varepsilon \pm \tilde{A}_3\varepsilon^{3/2} + \mathcal{O}(\varepsilon^2)$ with $A(\hat{\theta}) \neq 0$. For the second multiplier we start from identity (4.30) and apply l'Hopital's rule to $\partial_u\Theta = \frac{u^2-a(\varepsilon)}{\partial_\theta\hat{\phi}}$. This leads to

$$\left(\partial_u\Theta(\pm\sqrt{a(\varepsilon)},\varepsilon)\right)^2 = \frac{2u}{\partial_\theta^2\hat{\phi}}\bigg|_{u=\pm\sqrt{a(\varepsilon)}} = \frac{\pm2\sqrt{a(\varepsilon)}}{-\tilde{\omega}''(\hat{\theta}+\Delta\theta_\pm)} = \left(\frac{2}{\tilde{\omega}'''(\hat{\theta})}\right)^{2/3} + \mathcal{O}(\varepsilon^{1/2})\,.$$

Thus, we obtain

$$\alpha(\varepsilon) = A(\hat{\theta})\sqrt[3]{\tfrac{2}{|\tilde{\omega}'''(\hat{\theta})|}} + \mathcal{O}(\varepsilon)\,.$$

This identity again holds for $\varepsilon \in [-\varepsilon_0,\varepsilon_0]$. Note that the leading order term is always nonzero.

Concerning β we know by Theorem B.2 that the singularity in $\varepsilon = 0$ is removable. Indeed, we have $\beta = \partial_u f(0,0)$. Here we do not determine the leading order term explicitly. But note that in general this is not nonzero. For instance, in case $\hat{\phi}$ and with it U are symmetric (cf. special case of Theorem B.5) we have $\partial_u^2\Theta(0,0) = 0$. Thus for the first component of G (i.e. $A \equiv 1$) holds $\partial_u f(0,0) = A'(\hat{\theta})(\partial_u\Theta(0,0))^2 + A(\hat{\theta})\partial_u^2\Theta(0,0) = 0$.

Estimation of the error terms

Now we turn to estimate the error terms. We will prove that R_1 as well as R_2 and R_3 are at least $\mathcal{O}(t^{-1})$. To shorten the notion we introduce $u_\pm := U(\hat{\theta}\pm\delta_0,\varepsilon)$.

We start the considerations with R_2 where again partial integration does the trick,

$$\int\limits_{\mathbb{R}\setminus[u_-,u_+]} e^{it(u^3/3-a(\varepsilon)u)}du = \frac{1}{t\big[u^2-a(\varepsilon)\big]}\bigg|_{u_+}^{u_-} + \int\limits_{\mathbb{R}\setminus[u_-,u_+]} \frac{2u}{t\big[u^2-a(\varepsilon)\big]^2}e^{it(u^3/3-a(\varepsilon)u)}du\,.$$

Recall that $\left|\partial_\theta\hat{\phi}(\hat{\theta}\pm\delta_0,\varepsilon)\right| \geq \varepsilon_0$. Since $\partial_\theta\hat{\phi}\,\partial_u\Theta = u^2 - a(\varepsilon)$ and the fact that $|\partial_u\Theta|$ is uniformly bounded from below on $[\hat{\theta}-\delta,\hat{\theta}+\delta]\times[0,\varepsilon_0]$ we conclude $\left|u_\pm^2 - a(\varepsilon)\right| \geq C(\tilde{\omega})\varepsilon_0$. Obviously, the remaining integral on the right hand side also is uniformly bounded. Since the same arguments apply to $R_3(t,\varepsilon)$ we obtain

$$|R_2(t,\varepsilon)| \leq C(\tilde{\omega})\,t^{-1} \qquad \text{and} \qquad |R_3(t,\varepsilon)| \leq C(\tilde{\omega})\,t^{-1}$$

with $C(\tilde{\omega})$ only depending on $\tilde{\omega}$).

It remains to determine an upper bound on R_1. In view of (4.35) this is straight forward. We find

$$|R_1(t,\varepsilon)| \leq \big(|q(u_+,\varepsilon)| + |q(u_-,\varepsilon)| + \|\partial_u q(\cdot,\varepsilon)\|_{L^1(u_-,u_+)}\big)\, t^{-1} \leq C(\tilde{\omega})\, t^{-1}$$

which completes the proof.

We conclude the proof with two final remarks. Note first that one can not expect that the error bounds going to zero as $\varepsilon \to 0$. The reason is that R_2 and R_3, which represent the deviation from Airy's function due to the bounded domain of integration, as well as R_1, which represents the higher order terms of the asymptotic expansion, remain in case of a pointwise expansion for $\varepsilon = 0$. Second, the bounds $\sim t^{-1}$ of the error terms are again due to the cut-off. Using a smooth partition of unity R_2 and R_3 would behave like t^{-N} for all $N \in \mathbb{N}$ (cf. R_{loc}) and R_1 like the next term of the asymptotic expansion, namely $\sim t^{-4/3}$.

A Representations of Green's function

Recall the formulas of the dispersion relation,

$$\omega(\theta) := 2\left|\sin\tfrac{\theta}{2}\right|\omega_{\mathrm{r}}(\theta) \qquad \text{and} \qquad \tilde{\omega}(\theta) := 2\sin\tfrac{\theta}{2}\,\omega_{\mathrm{r}}(\theta)$$

with $\omega_{\mathrm{r}} \in \mathcal{C}^{\omega}(\mathcal{S}^1)$ and (2.12), i.e.

$$\begin{aligned} &\exists\, c_{\mathrm{r}} > 0\ \forall\, \theta \in \mathcal{S}^1 :\ \omega_{\mathrm{r}}(\theta) \geq c_{\mathrm{r}} \\ &\forall\, \theta \in \mathcal{S}^1 :\ \omega_{\mathrm{r}}(\theta) = \omega_{\mathrm{r}}(-\theta) = \omega_{\mathrm{r}}(\theta{+}2\pi). \end{aligned}$$

The following different representations of Green's function (2.8) are equivalent:

$$G_j(t) = \frac{1}{2\pi}\int\limits_{\mathcal{S}^1} \begin{pmatrix} \cos(\omega(\theta)t) & \frac{\mathrm{e}^{\mathrm{i}\theta}-1}{\omega(\theta)}\sin(\omega(\theta)t) \\ \frac{-\omega(\theta)}{\mathrm{e}^{\mathrm{i}\theta}-1}\sin(\omega(\theta)t) & \cos(\omega(\theta)t) \end{pmatrix} \mathrm{e}^{\mathrm{i}j\theta}\,\mathrm{d}\theta \tag{A.1}$$

$$G_j(t) = \frac{1}{2\pi}\int\limits_{\mathcal{S}^1} \begin{pmatrix} \cos(\tilde{\omega}(\theta)t) & \frac{\mathrm{i}\mathrm{e}^{\mathrm{i}\theta/2}}{\omega_{\mathrm{r}}(\theta)}\sin(\tilde{\omega}(\theta)t) \\ \mathrm{i}\mathrm{e}^{-\mathrm{i}\theta/2}\omega_{\mathrm{r}}(\theta)\sin(\tilde{\omega}(\theta)t) & \cos(\tilde{\omega}(\theta)t) \end{pmatrix} \mathrm{e}^{\mathrm{i}j\theta}\,\mathrm{d}\theta \tag{A.2}$$

$$G_j(t) = \frac{1}{2\pi}\int\limits_0^{\pi} \begin{pmatrix} h_+(\theta,t,j/t) & \frac{1}{\omega_{\mathrm{r}}(\theta)}h_-(\theta,t,(j{+}1/2)/t) \\ \omega_{\mathrm{r}}(\theta)h_-(\theta,t,(j{-}1/2)/t) & h_+(\theta,t,j/t) \end{pmatrix} \mathrm{d}\theta \tag{A.3}$$

where $\quad h_\pm(\theta,t,c) = \cos(t(\omega(\theta){+}\theta c)) \pm \cos(t(\omega(\theta){-}\theta c))$

$$G_j(t) = \frac{1}{2\pi}\int_{-\pi}^{\pi} \begin{pmatrix} \cos(\omega(\theta)t \pm \theta j) & \pm\frac{\operatorname{sign}(\theta)}{\omega_{\mathrm{r}}(\theta)} \cos(\omega(\theta)t \pm \theta(j+1/2)) \\ \pm\omega_{\mathrm{r}}(\theta) \operatorname{sign}(\theta) \cos(\omega(\theta)t \pm \theta(j-1/2)) & \cos(\omega_{\mathrm{r}} t \pm \theta j) \end{pmatrix} \mathrm{d}\theta \tag{A.4}$$

$$G_j(t) = \frac{1}{2\pi}\int_{S^1} \begin{pmatrix} \mathrm{e}^{\mathrm{i}t\phi_\pm(2\theta,c)} & \pm\frac{\mathrm{e}^{\mathrm{i}\theta}}{\omega_{\mathrm{r}}(2\theta)}\mathrm{e}^{\mathrm{i}t\phi_\pm(2\theta,c)} \\ \pm\frac{\omega_{\mathrm{r}}(2\theta)}{\mathrm{e}^{\mathrm{i}\theta}}\mathrm{e}^{\mathrm{i}t\phi_\pm(2\theta,c)} & \mathrm{e}^{\mathrm{i}t\phi_\pm(2\theta,c)} \end{pmatrix} \mathrm{d}\theta, \quad \phi_\pm(\theta,c) = \theta c \pm \tilde{\omega}(\theta). \tag{A.5}$$

Proof. We only give the proof for the two components $G_j^{1,1}(t)$ and $G_j^{1,2}(t)$. The proof for $G_j^{2,1}(t)$ is analogous to that of $G_j^{1,2}(t)$.

(A.1) $\Longrightarrow$ (A.2):
The result for the first component is obvious by $\cos(\omega(\theta)t) = \cos(\tilde{\omega}(\theta)t)$. For the second component we use $\sin(\omega(\theta)t) = \operatorname{sign}\theta \sin(\tilde{\omega}(\theta)t)$, and

$$2\sin\tfrac{\theta}{2} = -\mathrm{i}\mathrm{e}^{-\mathrm{i}\theta/2}(\mathrm{e}^{\mathrm{i}\theta} - 1) \text{ , i.e. } \quad \frac{\mathrm{e}^{\mathrm{i}\theta}-1}{\omega(\theta)} = \frac{\mathrm{i}\mathrm{e}^{\mathrm{i}\theta/2}}{\omega_{\mathrm{r}}(\theta)} \operatorname{sign}\theta \, .$$

(A.1) $\Longrightarrow$ (A.3):
For the first component we use

$$\int_0^{\pi} \cos(\omega t)\mathrm{e}^{\mathrm{i}\theta j}\,\mathrm{d}\theta = \frac{1}{2}\int_0^{\pi} \left[\mathrm{e}^{\mathrm{i}(\omega t+\theta j)} + \mathrm{e}^{-\mathrm{i}(\omega t-\theta j)}\right] \mathrm{d}\theta$$

$$\int_{-\pi}^{0} \cos(\omega t)\mathrm{e}^{\mathrm{i}\theta j}\,\mathrm{d}\theta = \frac{1}{2}\int_0^{\pi} \left[\mathrm{e}^{\mathrm{i}(\omega t-\theta j)} + \mathrm{e}^{-\mathrm{i}(\omega t+\theta j)}\right] \mathrm{d}\theta$$

and for the second

$$\begin{aligned}\int_0^{\pi} \frac{\mathrm{e}^{\mathrm{i}\theta}-1}{\omega}\sin(\omega t)\mathrm{e}^{\mathrm{i}\theta j}\,\mathrm{d}\theta &= \int_0^{\pi} \frac{\mathrm{i}\mathrm{e}^{\mathrm{i}\theta/2}}{\omega_{\mathrm{r}}} \cdot \frac{1}{2\mathrm{i}}\left[\mathrm{e}^{\mathrm{i}(\omega t+\theta j)} - \mathrm{e}^{-\mathrm{i}(\omega t-\theta j)}\right] \mathrm{d}\theta \\ &= \frac{1}{2}\int_0^{\pi} \frac{1}{\omega_{\mathrm{r}}}\left[\mathrm{e}^{\mathrm{i}(\omega t+\theta(j+1/2))} - \mathrm{e}^{-\mathrm{i}(\omega t-\theta(j+1/2))}\right] \mathrm{d}\theta\end{aligned}$$

$$\begin{aligned}\int_{-\pi}^{0} \frac{\mathrm{e}^{\mathrm{i}\theta}-1}{\omega}\sin(\omega t)\mathrm{e}^{\mathrm{i}\theta j}\,\mathrm{d}\theta &= \int_{-\pi}^{0} \frac{-\mathrm{i}\mathrm{e}^{\mathrm{i}\theta/2}}{\omega_{\mathrm{r}}} \cdot \frac{1}{2\mathrm{i}}\left[\mathrm{e}^{\mathrm{i}(\omega t+\theta j)} - \mathrm{e}^{-\mathrm{i}(\omega t-\theta j)}\right] \mathrm{d}\theta \\ &= \frac{1}{2}\int_0^{\pi} \frac{1}{\omega_{\mathrm{r}}}\left[\mathrm{e}^{-\mathrm{i}(\omega t+\theta(j+1/2))} - \mathrm{e}^{\mathrm{i}(\omega t-\theta(j+1/2))}\right] \mathrm{d}\theta \, .\end{aligned}$$

(A.3) $\Longrightarrow$ (A.4):
The proof for the first component is again straight forward since

$$\int_0^\pi \cos(\omega t \mp \theta j)\,\mathrm{d}\theta = \int_{-\pi}^0 \cos(\omega t \pm \theta j)\,\mathrm{d}\theta\,.$$

For the second component we have

$$\int_0^\pi \frac{1}{\omega_\mathrm{r}} \cos(\omega t \mp \theta(j+1/2))\,\mathrm{d}\theta = \int_{-\pi}^0 \frac{1}{\omega_\mathrm{r}} \cos(\omega t \pm \theta(j+1/2))\,\mathrm{d}\theta$$

and conclude

$$\int_0^\pi \frac{1}{\omega_\mathrm{r}} \Big[\cos(\omega(\theta)t+\theta(j+1/2)) - \cos(\omega(\theta)t-\theta(j+1/2))\Big]\,\mathrm{d}\theta$$

$$= \int_{-\pi}^\pi \frac{\mathrm{sign}(\theta)}{\omega_\mathrm{r}} \cos(\omega(\theta)t+\theta(j+1/2))\,\mathrm{d}\theta = -\int_{-\pi}^\pi \frac{\mathrm{sign}(\theta)}{\omega_\mathrm{r}} \cos(\omega(\theta)t-\theta(j+1/2))\,\mathrm{d}\theta\,.$$

(A.2) $\Longrightarrow$ (A.5):
Due to $\tilde{\omega}(2\pi+\theta) = -\tilde{\omega}(\theta)$ we have

$$\int_{-\pi}^\pi \mathrm{e}^{\mathrm{i}(\mp\tilde{\omega}t+\theta j)}\,\mathrm{d}\theta = \int_{-3\pi}^{-\pi} \mathrm{e}^{\mathrm{i}(\pm\tilde{\omega}t+\theta j)}\,\mathrm{d}\theta\,.$$

Thus, using the 4π-periodicity of $\tilde{\omega}$ we get

$$\int_{-\pi}^\pi \cos(\tilde{\omega}t)\mathrm{e}^{\mathrm{i}\theta j}\,\mathrm{d}\theta = \frac{1}{2}\int_{-\pi}^\pi \Big[\mathrm{e}^{\mathrm{i}(\tilde{\omega}t+\theta j)} + \mathrm{e}^{\mathrm{i}(-\tilde{\omega}t+\theta j)}\Big]\,\mathrm{d}\theta$$

$$= \frac{1}{2}\int_{-3\pi}^\pi \mathrm{e}^{\mathrm{i}(\pm\tilde{\omega}t+\theta j)}\,\mathrm{d}\theta = \int_{-\pi}^\pi \mathrm{e}^{\mathrm{i}(\pm\tilde{\omega}(2\theta)t+2\theta j)}\,\mathrm{d}\theta$$

which proves the result for the first component.
For the second component we use

$$\int_{-\pi}^\pi \frac{1}{\omega_\mathrm{r}} \mathrm{e}^{\mathrm{i}(\mp\tilde{\omega}t+\theta(j+1/2))}\,\mathrm{d}\theta = -\int_{-3\pi}^{-\pi} \frac{1}{\omega_\mathrm{r}} \mathrm{e}^{\mathrm{i}(\pm\tilde{\omega}t+\theta(j+1/2))}\,\mathrm{d}\theta$$

such that

$$\int_{-\pi}^{\pi} \frac{\mathrm{i}\mathrm{e}^{\mathrm{i}\theta/2}}{\omega_\mathrm{r}(\theta)} \sin(\tilde{\omega}(\theta)t)\mathrm{e}^{\mathrm{i}\theta j}\,\mathrm{d}\theta = \frac{1}{2}\int_{-\pi}^{\pi} \frac{1}{\omega_\mathrm{r}}\left[\mathrm{e}^{\mathrm{i}(\tilde{\omega}t+\theta(j+1/2))} - \mathrm{e}^{\mathrm{i}(-\tilde{\omega}t+\theta(j+1/2))}\right]\mathrm{d}\theta$$

$$= \pm\frac{1}{2}\int_{-3\pi}^{\pi} \frac{1}{\omega_\mathrm{r}}\mathrm{e}^{\mathrm{i}(\pm\tilde{\omega}t+\theta(j+1/2))}\,\mathrm{d}\theta = \pm\int_{-\pi}^{\pi} \frac{1}{\omega_{\mathrm{r}(2\theta)}}\mathrm{e}^{\mathrm{i}(\pm\tilde{\omega}(2\theta)t+2\theta(j+1/2))}\,\mathrm{d}\theta\,.$$

Finally, since $t\phi_\pm(\theta+2\pi, c) = t\phi_\pm(\theta, c) + 4\pi k$, the integrand is well defined on $\mathcal{S}^1$ and we actually have $\int_{\mathcal{S}^1} \dots \mathrm{d}\theta$ instead of $\int_{-\pi}^{\pi} \dots \mathrm{d}\theta$. □

Remark A.1:
We want to note the following observations.

- *In case of* $\omega_\mathrm{r} \equiv 1$ *the components of* $G_j(t)$ *are Bessel-functions (cf. [Fri03]).*
- ω *is* 2π-*, but* $\tilde{\omega}$ *is only* 4π*-periodic.*
- *For* $n \in \mathbb{N}$ *and* $k \in \mathbb{Z}$ *holds* $\omega^{(2n)}(0) = 0$, $\omega^{(2n+1)}(\pm\pi) = 0$ *and* $\tilde{\omega}^{(2n)}(2k\pi) = 0$.
- *The first term of* h *in* (A.3) *corresponds to wave numbers* $c \le 0$*, the second to* $c \ge 0$.
- *The kernel of* (A.4) *is only* 4π*-periodic, i.e. it is not* $\in C^\omega(S^1)$ *anymore and we really have* $\int_{-\pi}^{\pi} \dots \mathrm{d}\theta$ *instead of* $\int_{\mathcal{S}^1} \dots \mathrm{d}\theta$*. In contrast to that the kernel of* (A.5) *is* $C^\omega(S^1)$.
- *In passing to the representation* (A.5) *the number of critical wave numbers doubles: While the* sin- *and* cos-*terms for each wavenumber* θ *always involve two group velocities* $\pm\omega'(\theta)$*, the exponential terms necessitates two different wave numbers.*

B Preparation theorems

In Section 4 we derive uniform asymptotic expansions of integrals

$$I(t, y) = \int_{-\delta}^{\delta} A(\theta)\mathrm{e}^{\mathrm{i}t\phi(\theta,y)}\,\mathrm{d}\theta\,. \tag{B.1}$$

To do so we are faced with two challenges. First, it is necessary to rewrite the phase function ϕ in a suitable normal form, which is not straight forward if ϕ is degenerated. For instance in Section 4.3 we have $\phi(\theta, y)$ analytic with

$$\phi(0,0) = \partial_\theta\phi(0,0) = \partial_\theta^2\phi(0,0) = 0\,, \qquad \partial_\theta^3\phi(0,0) > 0 \tag{B.2}$$

but $\partial_\theta^2\phi(\theta, y) \neq 0$ for $(\theta, y) \neq 0$ in a neighborhood on $(0,0)$. Second, at one point we want to rewrite A in a suitable factorized form to be able to separate leading order terms, cf. (4.33) and (4.34). In both cases the key ingredients are the preparation theorems of Weierstrass and Malgrange, respectively.

To avoid confusion due to the non-uniform labeling we first recall the different versions of the preparation theorems according to [Hör90]. Then we state and proof a special version to be applied in Section 4.3.

Preparation theorems of Weierstrass and Malgrange

The first two theorems are the classical version in the analytic setting.

Theorem B.1 (Weierstrass preparation theorem, [Hör90, 7.5.1]):
Let g be an analytic function of $(\theta, z) \in \mathbb{C}^{1+n}$ in a neighborhood of $(0,0)$ such that

$$g(0,0) = \partial_\theta g(0,0) = \cdots = \partial_\theta^{k-1} g(0,0) = 0, \qquad \partial_\theta^k g(0,0) \neq 0 \tag{B.3}$$

Then there exists a unique factorization

$$g(\theta, z) = h(\theta, z)\left(\theta^k + a_{k-1}(z)\theta^{k-1} + \cdots + a_0(z)\right)$$

where a_j and h are analytic in a neighborhood of 0 and $(0,0)$ respectively, $h(0,0) \neq 0$ and $a_j(0) = 0$.

Sometimes the following division theorem is referred to as Weierstrass preparation theorem. It is a generalization of the last result and also known as Weierstrass formula.

Theorem B.2 (Weierstrass division theorem, [Hör90, 7.5.2]):
Let g and f be analytic functions of $(\theta, z) \in \mathbb{C}^{1+n}$ in a neighborhood of $(0,0)$ and g satisfy (B.3). Then

$$f(\theta, z) = q(\theta, z)g(\theta, z) + r_{k-1}(z)\theta^{k-1} + \cdots + r_0(z) \tag{B.4}$$

where r_j and q are uniquely determined and analytic in a neighborhood of 0 and $(0,0)$.

Theorem B.1 is a special case of Theorem B.2 with $f(\theta, z) = \theta^k$.

The C^∞ counterparts of the last two results are dedicated Malgrange. The analog of Theorem B.1, namely the Malgrange preparation theorem is again stated and proven in [Hör90]. Here we only state the second result which is sometimes also referred to as Malgrange preparation theorem or Mather division theorem.

Theorem B.3 (Malgrange division theorem, [Hör90, 7.5.6]):
Let g and f be C^∞ functions of $(\theta, x) \in \mathbb{R}^{1+n}$ in a neighborhood of $(0,0)$ and g satisfy (B.3). Then

$$f(\theta, x) = q(\theta, w)g(\theta, x) + r_{k-1}(x)\theta^{k-1} + \cdots + r_0(x)$$

where r_j and q are C^∞ functions in a neighborhood of 0 and $(0,0)$.

The proof can be deduced from the Weierstrass preparation theorem by decomposing a smooth function as a sum of analytic functions. But note that waiving the analyticity leads to a loss of uniqueness.

Essential in deriving a suitable normal for ϕ to rewrite (B.1) is that the normal form can also be achieved by a change of variables instead of a multiplication.

Theorem B.4 ([Hör90, 7.5.13]):
Let ϕ be a C^∞ function of $(\theta, y) \in \mathbb{R}^{1+n}$ in a neighborhood of $(0,0)$ which satisfies

$$\phi(0,0) = \partial_\theta \phi(0,0) = \cdots = \partial_\theta^{k-1}\phi(0,0) = 0, \qquad \partial_\theta^k \phi(0,0) > 0 \tag{B.5}$$

Then one can find a real valued C^∞ function $U(\theta, w)$ with $U(0,0) = 0$, $\partial_\theta U(0,0) > 0$ and C^∞ functions $a_j(y)$ with $a_j(0) = 0$ such that for $u = U(\theta, y)$ holds

$$\phi(\theta, y) = \tfrac{1}{k}u^k + a_{k-2}(y)u^{k-2} + \cdots + a_0(y) \,.$$

Real-analytic coordinate transform

In view of (B.2) we may apply Theorem B.4 to introduce a new coordinate $u = U(\theta, y)$ such that

$$\phi(\theta, y) = \tfrac{1}{3}u^3 + a(y)u + b(y) \,. \tag{B.6}$$

The functions U, a and b are C^∞ and $U(0,0) = a(0) = b(0) = 0$. But this result is insufficient in two ways. First it does not yield a real-analytic coordinate change. Second we claim to have $b(y) \equiv 0$ in a special case. Concerning this note that the 3rd power mixes even and odd powers of θ such that $U(-\theta, y) = -U(\theta, y)$, which would imply $b(y) \equiv 0$, it is not obvious. Here we give a modified version of Theorem B.4 which accounts for these two aspects.

Theorem B.5:
Let ϕ be real-analytic a function of $(\theta, y) \in \mathbb{R}^2$ in a neighborhood of $(0,0)$ which satisfies

$$\phi(0,0) = \partial_\theta \phi(0,0) = \cdots = \partial_\theta^{k-1}\phi(0,0) = 0, \qquad \partial_\theta^k \phi(0,0) > 0 \,. \tag{B.7}$$

Then one can find a real-analytic coordinate transform $u = U(\theta, y)$ and real-analytic functions $a_j(y)$ such that

$$\phi(\theta, y) = \tfrac{1}{k}u^k + a_{k-2}(y)u^{k-2} + \cdots + a_1(y)u + a_0(y) \tag{B.8}$$

Here $U(0,0) = 0$, $\partial_\theta U(0,0) > 0$ and $a_j(0) = 0$.

Furthermore if ϕ is odd w.r.t. θ, i.e. $\phi(-\theta, y) = -\phi(\theta, y)$, then so is U and $a_0(y) \equiv a_2(y) \equiv \cdots \equiv a_{k-3}(y) \equiv 0$.

Proof. For the general case the proof can be copied form that of Theorem B.4, see [Hör90, 7.5.13], using Weierstrass instead of Malgrange division theorem. Here we only give the proof for the specialization to the case where ϕ is odd.

Due to (B.7) there exists a function ϕ_0 such that $\phi(\theta, 0) = \frac{1}{k}\theta^k \phi_0(\theta)$ with $\phi_0(\theta) > 0$ in a sufficiently small neighborhood of 0. Now we may introduce the new coordinate $\zeta := \theta\phi_0^k(\theta)$, i.e. $\theta = \Theta(\zeta)$, such that $\phi(\Theta(\zeta), 0) = \frac{\zeta^k}{k}$ holds. Note that since ϕ odd w.r.t. θ, $k = 2J+1$ with $J \in \mathbb{N}$. We will use the notation $\tilde{\phi}(\zeta, y) := \phi(\Theta(\zeta), y)$. Here the symmetry of ϕ carries over to that of Θ and $\tilde{\phi}$ w.r.t. ζ.

We define

$$F(\zeta, y, \alpha) := \tilde{\phi}(\zeta, y) + \sum_{j=0}^{J-1} \alpha_j \zeta^{2j+1} .$$

Then we have in particular $F(\zeta, y, 0) = \phi(\Theta(\zeta), y)$. Now we want to let $\zeta = \zeta(y)$ and $\alpha = \alpha(y)$ vary such that $F(\zeta, y, \alpha)$ remains constant. Then $\frac{\mathrm{d}}{\mathrm{d}y}F = 0$ would lead to

$$\left(\frac{\partial \tilde{\phi}}{\partial \zeta} + \sum_{j=0}^{J-1}(2j+1)\alpha_j \zeta^{2j}\right)\frac{\mathrm{d}\zeta}{\mathrm{d}y} + \frac{\partial \tilde{\phi}}{\partial y} + \sum_{j=0}^{J-1}\frac{\mathrm{d}\alpha_j}{\mathrm{d}y}\zeta^{2j+1} = 0 . \tag{B.9}$$

Now we apply Weierstrass preparation theorem to justify the last equation. To do so we utilize the the symmetry of $\tilde{\phi}$. Note first that the function $\frac{\partial F}{\partial \zeta}$ is even w.r.t ζ. Thus

$$\Lambda(\xi, y, \alpha) := \left.\frac{\partial F}{\partial \zeta}\right|_{\zeta^2 = \xi} = \left.\frac{\partial \tilde{\phi}}{\partial \zeta}\right|_{\zeta^2=\xi} + \sum_{j=0}^{J-1}(2j+1)\alpha_j \xi^j$$

is real-analytic and satisfies (B.3) with $k = J$ and $w = (y, \alpha)$. For the same reason $\frac{\partial \tilde{\phi}}{\partial y}$ is odd w.r.t. ζ. Thus $f(\theta, y) := \left.\frac{1}{\zeta}\frac{\partial \tilde{\phi}}{\partial y}\right|_{\zeta^2=\xi}$ is again real-analytic. Applying Theorem B.2 to Λ and f leads to

$$\frac{1}{\zeta}\frac{\partial \tilde{\phi}}{\partial y} = q(\zeta^2, y, \alpha)\Big(\frac{\partial \tilde{\phi}}{\partial \zeta} + \sum_{j=0}^{J-1}(2j+1)\alpha_j\zeta^{2j}\Big) + \sum_{j=0}^{J-1} r_j(y, \alpha)\zeta^{2j} \tag{B.10}$$

In view of this relation (B.9) holds if

$$\frac{\mathrm{d}\zeta}{\mathrm{d}y} = -\zeta q(\zeta^2, y, \alpha) \qquad \text{and} \qquad \frac{\mathrm{d}\alpha_j}{\mathrm{d}y} = -r_j(y, \alpha) \quad \text{for} \quad j = 0, \ldots, J-1$$

does. Solving these ODE's with initial conditions $\zeta(0) = u$ and $\alpha_j(0) = a_j$ with u and a_j sufficiently close to 0 leads to $\zeta = \zeta(u, y, a)$ and $\alpha_j = \alpha_j(y, a_j)$. These functions are again real-analytic in a neighborhood of 0.

Since $\frac{\mathrm{d}}{\mathrm{d}y}F = 0$ in a neighborhood of 0 we conclude

$$\begin{aligned} F(\zeta, y, \alpha) &= \phi\big(\Theta(\zeta(u, y, a)), y\big) + \sum_{j=0}^{J-1} \alpha_j(y, a_j)[\zeta(u, y, a)]^{2j+1} \\ &= \phi\big(\Theta(\zeta(u, 0, a)), 0\big) + \sum_{j=0}^{J-1} \alpha_j(0, a_j)[\zeta(u, 0, a)]^{2j+1} \\ &= \frac{1}{k}u^k + \sum_{j=0}^{J-1} a_j u^{2j+1} \end{aligned}$$

Due to $\frac{\partial}{\partial a_j}\alpha_j(0, a_j) = 1$ and $\frac{\partial}{\partial u}\zeta(u, 0, a) = 1$ we can locally invert the functions $\alpha_j = \alpha_j(y, a_j)$ and $\zeta = \zeta(u, y, a)$. Hence $a_j = a_j(y, \alpha_j)$ and $u = u(\zeta, y, \alpha)$, where the dependence is again real-analytic. Thus we get

$$F(\zeta, y, \alpha) = \frac{1}{k}[u(\zeta, y, \alpha)]^k + \sum_{j=0}^{J-1} a_j(y, \alpha_j)[u(\zeta, y, \alpha)]^{2j+1} \,.$$

In view of the definition of F putting $\alpha = 0$ leads to the desired normal form and $U(\theta, y) := u(\Theta^{-1}(\theta), y, 0)$ is the corresponding coordinate transformation. Finally, the conditions on U and a_j in 0 as well as the symmetry of U w.r.t. θ are easily checked. □

References

[BH86] N. Bleistein and R. A. Handelsman. *Asymptotic expansions of integrals.* Dover Publications Inc., New York, second edition, 1986.

[FP99] G. Friesecke and R. L. Pego. Solitary waves on FPU lattices. I. Qualitative properties, renormalization and continuum limit. *Nonlinearity*, 12(6), 1601–1627, 1999.

[FPU55] E. Fermi, J. Pasta, and S. Ulam. Studies of nonlinear problems. *Los Alamos Scientific Laboratory of the University of California*, Report LA-1940, 1955.

[Fri03] G. Friesecke. Dynamics of the infinite harmonic chain: conversion of coherent initial data into synchronized binary oscillations. *Preprint*, 2003.

[GHM06] J. Giannoulis, M. Herrmann, and A. Mielke. Continuum descriptions for the dynamics in discrete lattices: derivation and justification. In *Analysis, modeling and simulation of multiscale problems*, pages 435–466. Springer, Berlin, 2006.

[HLTT08] L. Harris, J. Lukkarinen, S. Teufel, and F. Theil. Energy transport by acoustic modes of harmonic lattices. *SIAM J. Math. Anal.*, 40(4), 1392–1418, Jan. 2008.

[Hör90] L. Hörmander. *The analysis of linear partial differential operators. I*, volume 256 of *Grundlehren der Mathematischen Wissenschaften*. Springer-Verlag, Berlin, second edition, 1990.

[Ign07] L. I. Ignat. Fully discrete schemes for the Schrödinger equation. Dispersive properties. *Math. Models Methods Appl. Sci.*, 17(4), 567–591, 2007.

[IZ05] L. I. Ignat and E. Zuazua. Dispersive properties of a viscous numerical scheme for the Schrödinger equation. *C. R. Math. Acad. Sci. Paris*, 340(7), 529–534, 2005.

[Mie06] A. Mielke. Macroscopic behavior of microscopic oscillations in harmonic lattices via Wigner-Husimi transforms. *Arch. Ration. Mech. Anal.*, 181(3), 401–448, 2006.

[MP10] A. Mielke and C. Patz. Dispersive stability of infinite-dimansional Hamiltonian systems on lattices. *Applicable Analysis*, 89(9), 1493–1512, 2010.

[Olv74] F. W. J. Olver. *Asymptotics and special functions*. Academic Press, New York-London, 1974. Computer Science and Applied Mathematics.

[SK05] A. Stefanov and P. G. Kevrekidis. Asymptotic behaviour of small solutions for the discrete nonlinear Schrödinger and Klein-Gordon equations. *Nonlinearity*, 18(4), 1841–1857, 2005.

[Ste93] E. M. Stein. *Harmonic analysis: real-variable methods, orthogonality, and oscillatory integrals*, volume 43 of *Princeton Mathematical Series*. Princeton University Press, Princeton, NJ, 1993.

[Whi74] G. B. Whitham. *Linear and nonlinear waves*. Wiley-Interscience [John Wiley & Sons], New York, 1974. Pure and Applied Mathematics.

[Won89] R. Wong. *Asymptotic approximations of integrals*. Computer Science and Scientific Computing. Academic Press Inc., Boston, MA, 1989.

Chapter 4

Propagation of small amplitude wave fronts in infinite oscillator chains

The main result of Chapter 2 is that small localized initial data decay like in the linear case if the nonlinearity is of degree $\beta > 4$. On the other hand, for nonlinearities up to order $\beta < 3$ it is known that there exist solitary waves even for small and localized initial conditions. In this third paper, *Propagation of small amplitude wave fronts in infinite oscillator chains*, we aim to close this gap by numerical experiments and formal calculations. It turns out that in case $\beta = 3$ the wave fronts still decay like $\sim t^{-1/3}$, but this time the profile matches to the solution of a nonlinear ODE, namely a Painlevé equation. For $\beta \in (3, 4]$ there is a crossover between linear and nonlinear behavior at the front. Simulation results suggest that there are no solitary waves as long the initial conditions are sufficiently small.

Propagation of Small Amplitude Wave Fronts in Infinite Oscillator Chains

Carsten Patz

Humboldt-Universität zu Berlin, Institut für Mathematik
Rudower Chaussee 25, 12489 Berlin-Adlershof, Germany

30 September 2013

Contents

1 Introduction

This paper concerns long-time dynamics of Hamiltonian systems of interacting particles on one-dimensional infinite lattices. We study wave fronts generated by the evolution of small and localized initial conditions by formal calculations and numerical experiments. The aim is to understand the crossover between essentially linear and nonlinear behavior. In particular we analyze the order of the nonlinear part of the interaction (or background) potential at which wave fronts are dominated by the nonlinear and not any more by the linear part.

The prototype of Hamiltonian systems on lattices is the celebrated Fermi-Pasta-Ulam (FPU) chain

$$\ddot{x}_j(t) = V'(x_{j+1}-x_j) - V'(x_j-x_{j-1}) , \qquad j \in \mathbb{Z} , \tag{1.1}$$

where $V'(r) = ar + \mathcal{O}(r^\beta)$ as $r \to \infty$. This system exibits coherent structures first observed numerically by [FPU55]. These solitons were explained by deriving the completely integrable Korteweg-de Vries (KdV) equation in the long wave limit, cf. [ZK65]. In [FW94, FP99] the existence of solitary waves for generalized FPU systems is rigorously proven under additional global conditions on the interaction potentials V. Such waves satisfy $z_j(t) = Z(j-ct)$, where $z_j = (x_{j+1}-x_j, \dot{x}_j)$, for a fixed profile $Z : \mathbb{R} \to \mathbb{R}^2$ and a given wave speed c. In particular, [FW94] provides for the case $1 < \beta < 5$ the existence of solitary waves with arbitrarily small energy, i.e. $\|(\mathbf{x}^\delta_{\text{soli}}, \dot{\mathbf{x}}^\delta_{\text{soli}})\|_{\ell^2} = \delta \in (0, \delta_0)$.

Contrary to these results, in [GHM06, MP10] the autors address the question of dispersive stability of a more general class of systems including (1.1). That is, asking when solitary waves do not occur but solutions decay in time due to dispersive effects like it is the case in the linearized system, cf. [Fri03]. In fact, if $\beta > 4$ and the initial conditions are small and localized, i.e. small in l^p for $p < 2$, the solution to (1.1) decay $\sim t^{-1/3}$ and the leading order dynamics is driven by the linear part of V'. This implies that for $\beta > 4$ these solution cannot be small in ℓ^1.

In [FP99] the case $\beta = 2$ is investigated, and it is shown that $c = c_s + \mathcal{O}(\varepsilon^2)$, where c_s is the finite sonic wave speed of the linearized system. See also [McM02] and [FP02, FP04a, FP04b] in this context. The constructions there can be generalized to our case to provide small-energy solitary waves of associated with the generalized KdV limit. Moreover, in [SW00] it was shown that solutions of the form $r^\varepsilon_j(t) = \varepsilon^{2/(\beta-1)} R(\varepsilon^3 t, \varepsilon(j+\omega'(0)t)) + \text{h.o.t.}$ exist, with $R : [0, T] \times \mathbb{R} \to \mathbb{R}$ satisfying the generalized KdV equation

$$\partial_\tau R + b_1 \partial^3_\eta R + b_2 \partial_\eta V'(R) = 0 .$$

This equation possesses solitary wave solutions with exponentially decaying tales. In terms of the generalized FPU system these solutions satisfy $\|\mathbf{z}^\varepsilon_{\text{soli}}(t)\|_{\ell^1} \sim \varepsilon^{(3-\beta)/(\beta-1)}$. This shows that for $1 < \beta < 3$ there are solitary waves that are arbitrarily small in ℓ^1. We conclude that the above dispersive decay result cannot be transfered to $\beta < 3$. But the case $\beta \in [3,4]$ remains open.

This work is closely related to that field in a twofold way. First, and this is the actual motivation for this work, we aim to describe the gap $\beta \in [3,4]$, at least by formal calculations and numerical experiments. We expect the stability result to be not optimal. In particular it is an open question whether the approach of treating the nonlinearity as a small perturbation of the linear part is exhausted. Now, in l^p, $p > 4$, the decay of the linearized system is dominated by the region near the wave fronts, cf. again [MP10]. Here we analyzes the relation between the order of the nonlinearity and the question whether wave fronts emerging by the evolution of localized initial conditions are dominated by linear or nonlinear dynamics. It turns out that in case $\beta = 3$ the wavefronts still decay $\sim t^{-1/3}$ like in, but this time the front matches to the solution of a nonlinear ODE, namely a Painlevé equation. In that case we claim that perturbation arguments will fail and a substantial nonlinear theory is necessary. But numerical experiments suggest that the solutions are still asymptotically stable if the initial data are sufficiently small. For $\beta \in (3,4)$ the simulations show a behavior that is somehow in between the Airy- and the Painlevé-front.

Second, it turns out that the wave fronts considered also lie in the long-wave-limit regime and one expects a behavior similar to a KdV-type equation. We also investigate the question whether or in what sense such an approximation is reasonable.

In fact we study three special cases of a more general system including (1.1). We consider an infinite number of equal particles with unit mass interacting with a finite number K of neighbors via potentials $V_1, \ldots, V_K$ and possibly coupled to a background via an on-site potential W. According to Newton's law the equations of motion are

$$\ddot{x}_j = \sum_{1 \le k \le K} \Big(V_k'(x_{j+k} - x_j) - V_k'(x_j - x_{j-k}) \Big) - W'(x_j) \,, \qquad j \in \mathbb{Z}. \tag{1.2}$$

Here $x_j \in \mathbb{R}$ denote the displacements. We will use the notation $\mathbf{x} := (x_j)_{j \in \mathbb{Z}}$ and assume that for the interaction and the on-site potentials are smooth and

$$V_k'(r) = a_k r + V_{\mathrm{nl},k}'(r) \qquad \text{with} \quad V_{\mathrm{nl},k}'(r) = \mathcal{O}(|r|^\beta)_{|r| \to 0}.$$

On $W'(x)$ we impose similar assumptions. For the numerical studies we will restrict ourselves to prototypical nonlinearities summarized by $\mathcal{N}_j(\mathbf{x})$ consisting of terms $|y|^\beta$ or $y|y|^{\beta-1}$ for $y = x_j$ and $y = x_{j+1} - x_j$, respectively, such that

$$\mathcal{N}_j(\varepsilon \mathbf{x}) = \varepsilon^\beta \mathcal{N}_j(\mathbf{x}) \tag{1.3}$$

holds for all $\varepsilon > 0$.

System (1.2) is Hamiltonian, i.e. $(\dot{\mathbf{x}}, \dot{\mathbf{p}})^T = \mathcal{J}_{\mathrm{can}} \, \mathrm{d}\mathcal{H}_{\mathbf{x}}(\dot{\mathbf{x}}, \dot{\mathbf{p}})$ with momentum $\mathbf{p} := \dot{\mathbf{x}}$, $\mathcal{J}_{\mathrm{can}}$ the Poisson tensor corresponding to the canonical symplectic structure defined by $\langle(\mathbf{x},\mathbf{p}),\mathcal{J}_{\mathrm{can}}(\tilde{\mathbf{x}},\tilde{\mathbf{p}})\rangle_{\ell^2\oplus\ell^2}=\langle\mathbf{x},\tilde{\mathbf{p}}\rangle_{\ell^2}-\langle\tilde{\mathbf{x}},\mathbf{p}\rangle_{\ell^2}$ and total energy $\mathcal{H}_{\mathrm{x}}(\mathbf{x},\mathbf{p})=\sum_{j\in\mathbb{Z}}\left(\frac{1}{2}p_j^2+\sum_{1\le k\le K}V_k(x_{j+k}-x_j)+W(x_j)\right)$.

The dispersive decay is driven by the linearized system

$$\ddot{x}_j=-\mathcal{A}_j\,\mathbf{x}\,,\qquad \text{where}\quad \mathcal{A}_j\,\mathbf{x}:=\sum_{-K\le k\le K}a_kx_j \tag{1.4}$$

with $a_{-k}=a_k$ and $a_0=-2\sum_{1\le k\le K}a_k+b$. The dispersion relation is obtained by looking for plane waves in the form $x_j(t)=\mathrm{e}^{\mathrm{i}(\theta j+\tilde{\omega}t)}$. We find the relation

$$\omega^2=\mathbb{A}(\theta):=\sum_{-K\le k\le K}a_k\mathrm{e}^{\mathrm{i}k\theta}=a_0+2\sum_{1\le k\le K}a_k\cos(k\theta). \tag{1.5}$$

Here $\mathbb{A}(\theta)$ is the symbol of $\mathcal{A}$. Throughout, we make the stability condition

$$\mathbb{A}(\theta)>0\quad\text{for all }\theta\in\mathcal{S}^1\setminus\{0\}.$$

Thus we may define $\omega(\theta):=\sqrt{\mathbb{A}(\theta)}$ which is smooth at least on $\mathcal{S}^1\setminus\{0\}$.

We want to study the evolution of wave fronts of solutions of (1.2). On the level of a formal calculation this is possible by considering the lattice system in the long wave limit to derive a PDE, cf. [GM06, GHM06], and study its self-similar solutions. To do so we fix some wave number θ_0 with $\omega''(\theta_0)=0$, i.e. θ_0 is a critical wave number corresponding to two wave fronts of the linearized system. To resolve the wave fronts we use the *KdV-* or *Airy-scaling* to introduce new macroscopic variables

$$\tau:=\varepsilon^3t\,,\qquad \eta:=\varepsilon(j-ct)\,, \tag{1.6}$$

where $c:=\omega'(\theta_0)$ and consider solutions of (1.2) of the form

$$\begin{aligned}x_j(t)&=\sum_{l=1}^{L}\varepsilon^{\gamma_l}\sum_{n=-l}^{l}A_{l,n}\left(\varepsilon^3t,\varepsilon(j-ct)\right)\mathrm{e}^{\mathrm{i}n(\omega(\theta_0)t-j\theta_0)}+o(\varepsilon^{\gamma_L})\\&:=X_j^{\varepsilon}(t)+o(\varepsilon^{\gamma_L})\end{aligned} \tag{1.7}$$

for $j\in\mathbb{Z}$, where $A_{l,n}\in\mathbb{C}$, $\gamma_l\in\mathbb{R}_0^+$ and $L\in\mathbb{N}$ fixed. To obtain a real-valued solution we require $A_{l,n}=\bar{A}_{l,-n}$ which implies $A_{l,0}\in\mathbb{R}$. We will use the notations

$$E_j(t):=\mathrm{e}^{\mathrm{i}(\omega(\theta_0)t-j\theta_0)}\,,\qquad X_{l,n,j}^{\varepsilon}(t):=\varepsilon^{\gamma_l}A_{l,n}\left(\varepsilon^3t,\varepsilon(j-ct)\right)E_j^n(t)+\text{c.c.}\,,$$

$\mathbf{X}^{\varepsilon}:=(X_j^{\varepsilon})_{j\in\mathbb{Z}}$ and $\mathbf{X}_{l,n}^{\varepsilon}:=(X_{l,n,j}^{\varepsilon})_{j\in\mathbb{Z}}$, where "c.c." stands for the complex conjugate if the imaginary part is non-zero. The "$-$" in the exponent in (1.7) accounts for the fact that the modulating wave E travels in the same direction like the enveloping profile $A_{l,n}$.

For FPU-type systems, that is systems like (1.1) where the restoring force $-\frac{\partial}{\partial x_j}\mathcal{V}(\mathbf{x})$ only represents an interaction force between the particles, the modulation terms $E_j^n(t)$ in (1.7) cancels in case we consider the the front traveling the sonic wave speed. Thus this ansatz reduces to the well known connection between FPU and KdV or generalized KdV (gKdV). We study this case in detail in Section 3. Otherwise the modulation term is necessary. Hence we also need the higher order terms since the nonlinearity produces higher order harmonics, i.e. terms $E_j^n(t)$. Here even the formal calculations are not straight forward. We discuss one example in Appendix B.

In any case we end up with a PDE for the leading order term $A(\tau,\eta) := A_{l,n}(\tau,\eta)$. Assuming the wave front to have a self-similar structure leads to the ansatz

$$A(\tau,\eta) := \tau^{\tilde{\gamma}}\phi\Big(\frac{\eta}{\tau^{1/3}}\Big)\,, \quad \xi := \frac{\eta}{\tau^{1/3}}\,, \qquad \text{where} \quad \tilde{\gamma} = -\frac{\gamma}{3}\,. \tag{1.8}$$

The main question is, for which β does the profile ϕ reasonably approximate the long-time behavior of the wave front in the actual lattice system.

To answer this question we numerically study the evolution of small localized initial conditions in systems of the form (1.2) for different β on the time and space scales indicated by the formal analysis. We restrict ourselves to two prototypes of localized initial conditions, namely the Riemann problem and the δ-peak. The smallness of the initial conditions will, in view of (1.3), be taken into account by a factor in front of the nonlinearity. The localization is motivated by our aim to examine the decay of microscopic initial data. While this also avoids the occurrence of KdV-like solitary waves which are small in amplitude but scale like $\frac{1}{\varepsilon}$ in space, see [FP99], the smallness of the initial data is to suppress high energy solitary waves, see [RH93, FM02]. The latter is what we generically see in simulations if the initial data are not small enough.Since solitary waves are supersonic we do not claim that the occurrence of solitary waves significantly changes the qualitative asymptotic behavior of the wave fronts under consideration.

This paper is organized as follows. First, in the upcoming section 2, we discuss the linear case. Here we collect results which seem to be known but are scattered in the literature. On the other hand we have the feeling that wave propagation in discrete lattices is not as general knowledge as the corresponding PDE theory. In this manner this sections serves as an introduction to the topic. In section 3 we treat FPU-type systems with nonlinear interaction potentials. Finally, in section B consider systems with nonlinear background potentials.

2 Linear systems

First we treat the linear case, i.e.

$$\ddot{x}_j = -\mathcal{A}_j\mathbf{x}\,, \qquad j \in \mathbb{Z}\,, \tag{2.1}$$

where $\mathcal{A}$ is defined in (1.4). This system can be solved explicitly using Fourier transform, cf. [MP10]. We aim to resolute the wave front corresponding to some wave number θ_0 characterized by $\omega''(\theta_0) = 0$ traveling with speed $c = \omega'(\theta_0) > 0$.

To exclude degenerated cases we impose one further condition on the dispersion relation, namely

$$\omega'''(\theta_0) \neq 0 \,. \tag{2.2}$$

Then the method of stationary phase states that the wave front decays like $\sim t^{-1/3}$ while the generic decay is $\sim t^{-1/2}$, see e.g. [Won89] for the general approach and [Fri03, MP10] for details concerning lattice systems. It turns out that rescaling these fronts basically leads to an Airy function. The rigorous proof, cf. [BH86, Won89, Hör90, MP13], uses a parameter-dependent version of the method of stationary phase. Thus it is restricted to the linear case. Here we retrace the result by a formal analysis also to be applied to nonlinear systems in the upcoming sections.

In fact we have to distinguish two different cases. First, we have FPU-type systems that allow for a formulation in terms of distances $r_j = x_{j+1} - x_j$ due to Galilean invariance. The equation of motion in terms of $\mathbf{r}$ again reads like (2.1). In this case the wave number $\theta_0 = 0$ stands out and the corresponding wave front bears no oscillation term. Second we consider the case with modulation term, in particular systems with on-site potential (KG-type systems).

Figure 2.1 and 2.2 shows the dispersion relations and time evolutions for the three prototypical cases, namely FPU-type systems with nearest-neighbor (NN) interaction and nearest- and next-nearest-neighbor (NNN) interaction and a KG-type system. The parameters in the NNN-case are chosen such that the wave speeds of the fronts are ± 1 and ± 0.5, respectively.

2.1 Monotone Airy-fronts in FPU-type systems

In case of FPU-type systems for the dispersion relation holds

$$\omega(\theta) = \sqrt{\mathbb{A}(\theta)} = 2\left|\sin\tfrac{\theta}{2}\right| \omega_{\mathbf{r}}(\theta) \,, \qquad \forall\, \theta \in S^1 : \omega_{\mathbf{r}}(\theta) = \omega_{\mathbf{r}}(-\theta) > 0 \,, \tag{2.3}$$

where $\omega_{\mathbf{r}}$ is a smooth function, cf. [MP10, MP13]. Hence the wave number $\theta_0 = 0$ is always critical in the sense that $w''(0) = 0$) and due to $\omega(0) = 0$ it bears no oscillations, i.e. the modulation term in (1.7) becomes obsolete. If $c := |\omega'(0)| = \max_{\theta \in S^1 \setminus} |\omega'(\theta)|$ FPU-type system shape two locally monotone fronts corresponding to $\theta_0 = 0$.

Since in the linear case the terms do not interact it is sufficient to use the ansatz

$$X_j^\varepsilon(t) = A\left(\varepsilon^3 t, \varepsilon(j - ct)\right) \,, \quad j \in \mathbb{Z} \tag{2.4}$$

instead of (1.7). Employing this ansatz in (2.1) the terms corresponding to ε^2 cancel due to $\partial_\theta^2 \mathbb{A}(0) = 2(\omega'(0))^2 = 2c^2$. The condition on the lowest remaining order terms ε^4 reads as

$$\partial_{\tau\eta} A + \frac{a}{3}\partial_\eta^4 A = 0 \,, \qquad a = -\frac{\partial_\theta^4 \mathbb{A}(0)}{16c} \,. \tag{2.5}$$

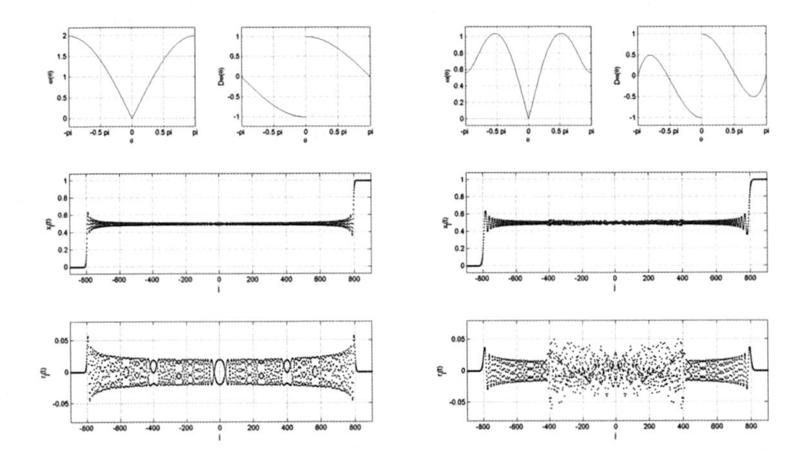

Figure 2.1: Dispersion relations and time evolutions for FPU-NN ($a_1 = -1$) and FPU-NNN ($a_1 = -0.08$, $a_2 = -0.23$) with two wave fronts: $\omega(\theta)$, $\omega'(\theta)$, $x_j(t)$ and $r_j(t)$, respectively, at $t = 800$ to initial condition $(r_j(0), \dot{r}_j(0)) = (\delta_{j,0}, 0)$.

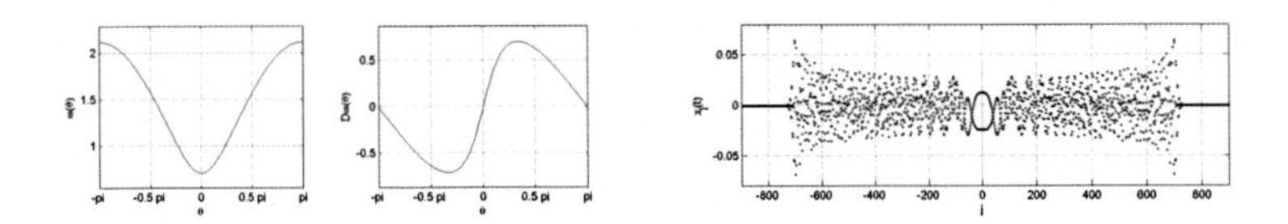

Figure 2.2: Dispersion relation and time evolution for the prototypical KG chain ($a_1 = -1$, $b = 0.5$): $\omega(\theta)$, $\omega'(\theta)$ and $x_j(t)$ at $t = 800$ to initial condition $(x_j(0), \dot{x}_j(0)) = (\delta_{j,0}, 0)$.

This is Airy's PDE in terms of $\partial\eta A$. Note that (2.5) is again Hamiltonian with $\mathcal{H}(A) = \frac{\partial_\theta^4 \mathbb{A}(0)}{48} \int_{\mathbb{R}} (\partial_\eta^2 A)^2 \, \mathrm{d}\eta$ and symplectic structure $\mathbb{S}^{\mathrm{red}} = 2c\partial_\eta$, i.e. $\mathbb{S}^{\mathrm{red}}\partial_\tau A = \mathrm{D}\mathcal{H}(A)$, cf. (A.8).

We expect the wave fronts to converge a self-similar structure. Employing the ansatz (1.8) in (2.5) leads to

$$(\gamma - \tfrac{1}{3})\phi' - \tfrac{1}{3}\xi\phi'' + \tfrac{a}{3}\phi^{(4)} = 0 \, .$$

Assuming sufficiently fast decay of $\phi(\xi)$ and its derivatives as $\xi \to +\infty$ integration yields

$$\tilde{\gamma}\phi - \tfrac{1}{3}\xi\phi' + \tfrac{a}{3}\phi''' = 0 \, . \tag{2.6}$$

Now we have to distinguish two cases corresponding to amplitudes and relative distances, respectively. Note first that $\tilde{\gamma}$ corresponds to the decay rate in time of the wave crest. Assuming initial conditions in terms of relative distances $(\mathbf{r}^0, \mathbf{p}^0) \in l^1(\mathbb{Z}, \mathbb{R}^2)$, which is natural from the physical point of view, we can not expect any decay in terms of absolute displacements. That is, in this case we have $\tilde{\gamma} = 0$, cf. Figure 2.1. But in terms of distances the solution near the wave front decays like $t^{-1/3}$. Thus we have $\tilde{\gamma} = -\frac{1}{3}$.

Now, setting $\tilde{\gamma} = 0$ in (2.6) and writing $\phi(\xi) = \int_{\infty}^{\xi} \hat{\phi}(\zeta)\,\mathrm{d}\zeta + x_\infty$ leads to $\hat{\phi}'' = \frac{1}{a}\,\xi\hat{\phi}$. In case $\tilde{\gamma} = -\frac{1}{3}$ we may integrate (2.6) to obtain once again

$$\phi'' = \tfrac{1}{a}\,\xi\phi\,. \tag{2.7}$$

Thus in both cases we end up with a rescaled Airy equation describing the profile of the wave front. As it is to be expected the profiles in terms of distances and amplitudes are linked by an integration.

Airy's equation $\phi'' = \xi\phi$ admits two linear independent solutions. One of those decays monotonic and super-exponentially as $\xi \to \infty$, namely the Airy function

$$\mathrm{Ai}(\xi) = \frac{1}{2\pi} \int_{\mathbb{R}} \mathrm{e}^{\mathrm{i}(1/3z^3 + \xi z)}\,\mathrm{d}z\,.$$

Then the suitable solution of (2.7) reads as

$$\phi(\xi) = a^{-1/3} A_0 \mathrm{Ai}\big(a^{-1/3}\xi\big)\,, \tag{2.8}$$

where A_0 is to be determined by the initial conditions.

We determine A_0 from the initial data of the original lattice system (2.1) in terms of r_j. We claim that

$$r_j(t)\Big|_{j=ct} \sim \phi(0)\,t^{-1/3}\,, \qquad \text{as} \quad t \to \infty \tag{2.9}$$

The asymptotic behavior of the solution of the lattice system can be determined using the method of stationary phase.

Proposition 2.1:
Consider the system (2.1) *with dispersion relation* (2.3)*. Assume that* (2.2) *holds for* $\theta_0 = 0$*. Then the unique solution for initial data* $(\mathbf{r}^0, \mathbf{p}^0) \in l^1(\mathbb{Z}, \mathbb{R}^2)$ *satisfies*

$$\begin{pmatrix} r_j(t) \\ p_j(t) \end{pmatrix}\Bigg|_{j=\pm ct} = \frac{c_*}{|\omega'''(0)|^{1/3}} \sum_{k\in\mathbb{Z}} \begin{pmatrix} 1 & \mp\frac{1}{c} \\ \mp c & 1 \end{pmatrix} \begin{pmatrix} r_k^0 \\ p_k^0 \end{pmatrix} t^{-1/3} + \mathcal{O}(t^{-2/3}) \tag{2.10}$$

with $c_* = \frac{\Gamma(1/3)}{2^{5/3}3^{1/6}\pi}$.

We skip the proof since the result is straight forward by applying the method, for instance according to [Won89] to the explicit solution of (2.1) with (2.3), cf.

[Fri03]. Or, it also might be deduced from the more general result [MP13, Thm 3.3].

Now, combining (2.10) with (2.9) we obtain the condition $c_*/|\omega'''(0)|^{1/3}\sum_{k\in\mathbb{Z}}\big(r_k^0 \mp \frac{1}{c}p_k^0\big) = A_0 a^{-1/3}\mathrm{Ai}(0)$. Using $\mathrm{Ai}(0) = 1/3^{2/3}\Gamma(2/3)$ and $\Gamma(1-z)\Gamma(z) = \pi/\sin\pi z$ we get $A_0 = 1/2\sum_{k\in\mathbb{Z}}\big(r_k^0 \mp 1/c\,p_k^0\big)$. Thus we end up with

$$\phi(\xi) = \frac{1}{2a^{1/3}}\sum_{k\in\mathbb{Z}}\big(r_k^0 \mp \tfrac{1}{c}p_k^0\big)\cdot \mathrm{Ai}\Big(\frac{\xi}{a^{1/3}}\Big)\,, \qquad a = \frac{|\omega'''(0)|}{2}\,. \tag{2.11}$$

Figure 2.3 shows simulation results for the NN-case. We see a perfect match of the scaled solution of $\ddot{r}_j = r_{j-1} - 2r_j + r_{j+1}$ to the initial data $(\mathbf{r}^0, \dot{\mathbf{r}}^0) = ((\delta_{j,0})_{j\in\mathbb{Z}}, \mathbf{0})$, i.e. $t^{1/3}r_j(t)$ versus $\xi = (j-ct)/t^{1/3}$ with the Airy function.

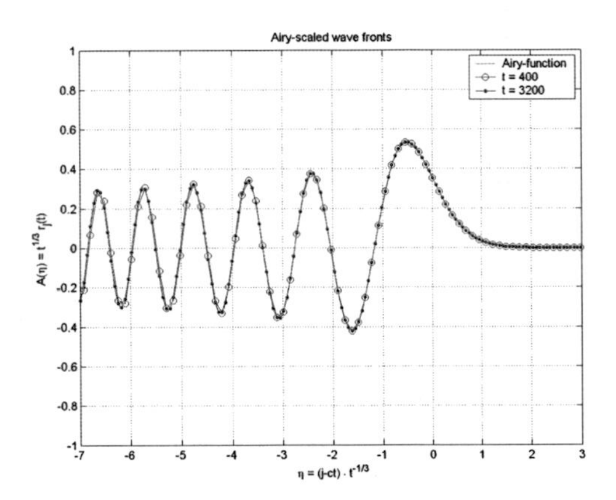

Figure 2.3: Scaled wave front (FPU NN): $\phi(\xi) = t^{1/3}r_j(t)$ vs. $\eta = t^{-1/3}(j-t)$ and Airy-profil according to (2.11).

We conclude this section with three remarks. Note first that the condition (2.2) is essential for a decay rate $-1/3$. For instance, $\omega'''(\theta_0) = 0$ but $\omega^{(4)}(\theta_0) \neq 0$ would imply a decay rate of $-1/4$. Note second that, since the width of the front scales like $t^{1/3}$ and its height like $t^{-1/3}$, the front conserves the l^1-norm. The same argument shows that the energy at the front tends to 0 as $t \to \infty$. Thus the whole energy is contained in the inner region decaying like $t^{-1/2}$, cf. [Mie06], in particular Fig. 6.2., in this context.

2.2 Oscillatory Airy-fronts in KG- and FPU-type systems

It remains the case of critical wave numbers $\theta_0 \neq 0$ which is to be treated in the same way like systems with on-site potential. For the latter, due to $\mathbb{A}(\theta) > 0$ for all $\theta \in S^1$, all particles within the lightcone always oscillate. Thus wave front is a profile modulated by $E_j(t) = \mathrm{e}^{\mathrm{i}(\omega(\theta_0)t - \theta_0 j)}$. Employing the ansatz

$$X_j^\varepsilon(t) = \varepsilon^\gamma A\left(\varepsilon^3 t, \varepsilon(j - ct)\right) \mathrm{e}^{\mathrm{i}(\omega(\theta_0)t - \theta_0 j)}\ , \quad j \in \mathbb{Z} \tag{2.12}$$

in (2.1) this time leads to

$$\partial_\tau A + \tfrac{a}{3}\, \partial_\eta^3 A = 0\ , \qquad a = -\frac{\partial_\theta^3 \mathbb{A}(\theta_0)}{4\omega(\theta_0)}\ . \tag{2.13}$$

Note that this is (2.5) once integrated with respect to η. Thus, this time employing (1.8) immediately leads to (2.7). Here $\tilde{\gamma} = 0$ corresponds to FPU-type systems in terms of x_j and $\tilde{\gamma} = -\frac{1}{3}$ to KG-type systems in terms of x_j or FPU-type systems in terms of r_j.

Figure 2.4 and 2.5 show simulation results for the KG-type case. The first plot shows the scaled solution of $\ddot{x}_j = x_{j-1} - 2.5x_j + x_{j+1}$, i.e. $t^{1/3} r_j(t)$, to the initial data $(\mathbf{x}^0, \dot{\mathbf{x}}^0) = \left((\delta_{j,0})_{j\in\mathbb{Z}}, \mathbf{0}\right)$ and the modulated Airy profile

$$\Phi(\eta) = \cos\left(\omega(\theta_0)t - \theta_0 j\right) A_0 \mathrm{Ai}\left(\frac{j - ct}{a^{1/3} t^{1/3}}\right)\ .$$

versus $\xi = \frac{j-ct}{t^{1/3}}$.

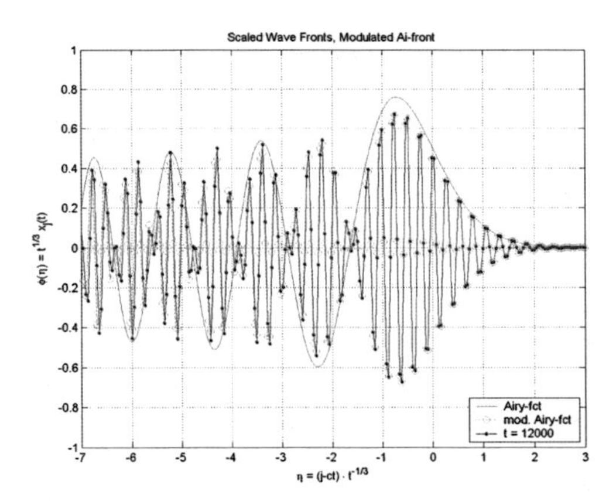

Figure 2.4: Scaled wave fronts of KG chain: $\phi(\xi)$, Airy function and modulated Airy-profile vs. ξ.

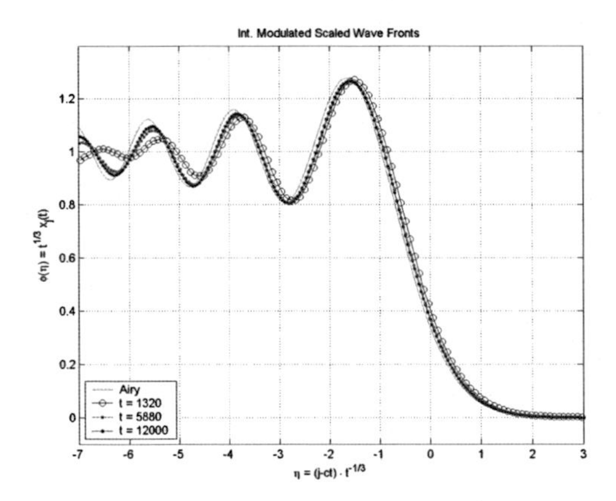

Figure 2.5: Wave fronts of the KG chain: $\int_{\infty}^{\eta} Re(A(\tilde{\eta})E_j(t))\, \mathrm{d}\tilde{\eta}$ vs. η and Airy function.

3 Nonlinear FPU-type systems

3.1 Formal Derivation

We consider a system given by

$$\ddot{x}_j = -\mathcal{A}_j\,\mathbf{x} + V'_{\mathrm{nl}}(x_{j+1}-x_j) - V'_{\mathrm{nl}}(x_j - x_{j-1})\,, \quad j\in\mathbb{Z}\,, \tag{3.1}$$

or in terms of relative displacements

$$\ddot{r}_j = -\mathcal{A}_j\,\mathbf{r} + V'_{\mathrm{nl}}(r_{j+1}) - 2V'_{nl}(r_j) + V'_{\mathrm{nl}}(r_{j-1})\,, \quad j\in\mathbb{Z}\,. \tag{3.2}$$

The nonlinear part V_{nl} of the interaction potential is assumed to be smooth and $V'_{\mathrm{nl}}(r) = \mathcal{O}(|r|^\beta)$. For simplicity we only consider

$$V_{\mathrm{nl}}(r) = \frac{1}{\beta+1}|r|^{\beta+1}\,, \qquad \text{or} \qquad V_{\mathrm{nl}}(r) = \frac{1}{\beta+1}r|r|^\beta \tag{3.3}$$

for $\beta \geq 2$. The higher order terms will cancel out anyway by envolving Taylor series.

First we consider (3.1). Due to $\omega(0)=0$ for the modulation term in (1.7) holds $E=1$, i.e. the modulation term becomes obsolete. Furthermore it turns out that the ansatz

$$x_j(t) = X^\varepsilon_j(t) + \mathcal{O}(\varepsilon^{\gamma+1}) = \varepsilon^\gamma A\left(\varepsilon^3 t, \varepsilon(j-ct)\right) + \mathcal{O}(\varepsilon^{\gamma+1}) \tag{3.4}$$

for $j\in\mathbb{Z}$ is sufficient which will become clear by the expansion of the nonlinear terms. Using $\partial_t^2 = \varepsilon^6\partial_\tau^2 - 2c\,\varepsilon^4\partial_{\tau\eta} + c^2\varepsilon^2\partial_\eta^2$ and Corollary A.3 we obtain for the linear terms

$$(\partial_t^2 + \mathcal{A})_j\mathbf{X}^\varepsilon = \left(-2c\,\partial_{\tau\eta}A + \frac{\partial_\theta^4\mathbb{A}(0)}{24}\,\partial_\eta^4 A\right)\varepsilon^{4+\gamma} + \mathcal{O}(\varepsilon^{5+\gamma})\,.$$

Concerning the nonlinear terms Taylor expansion leads to

$$V'_{nl}\big(X^\varepsilon_{j+1} - X^\varepsilon_j\big) - V'_{nl}\big(X^\varepsilon_j - X^\varepsilon_{j-1}\big)\big) = \varepsilon^{\beta\gamma+\beta+1}\partial_\eta V'_{nl}(\partial_\eta A) + \mathcal{O}(\varepsilon^{\beta\gamma+\beta+2})\,.$$

Balancing linear and nonlinear terms leads to $\gamma+4 = \beta\gamma+\beta+1$. Thus, if V_{nl} is smooth the system admits solutions of the form (3.4) for $\gamma = -\frac{\beta-3}{\beta-1}$ and all $\varepsilon\in(0,\varepsilon_0)$ for some $\varepsilon_0>0$, then A satisfies the PDE

$$2c\,\partial_{\tau\eta}A - \frac{\partial_\theta^4\mathbb{A}(0)}{24}\,\partial_\eta^4 A + \partial_\eta V'_{nl}(\partial_\eta A) = 0\,. \tag{3.5}$$

The ansatz

$$A(\tau,\eta) := \tau^\gamma\phi\Big(\frac{\eta}{\tau^{1/3}}\Big)\,, \qquad \xi := \frac{\eta}{\tau^{1/3}} \tag{3.6}$$

yields

$$\partial_{\tau\eta} A = (\gamma - \frac{1}{3})\tau^{\gamma-4/3}\phi' - \frac{1}{3}\tau^{\gamma-4/3}\xi\phi'' , \qquad \partial_\eta^4 A = \tau^{\gamma-4/3}\phi^{(4)}$$

$$\partial_\eta(\partial_\eta A)^n = \tau^{\beta\gamma-(\beta+1)/3}\frac{\mathrm{d}}{\mathrm{d}\xi}V'_{nl}\big(\phi'(\xi)\big) .$$

Employing in (3.5) and using $\gamma_\beta = \frac{\beta-3}{3(\beta-1)}$ leads to

$$2c\Big((\gamma_n - \frac{1}{3})\phi'(\xi) - \frac{1}{3}\xi\phi''(\xi)\Big) - \frac{\partial_\theta^4\mathbb{A}(0)}{24}\phi^{(4)}(\xi) + \frac{\mathrm{d}}{\mathrm{d}\xi}V'_{nl}\big(\phi'(\xi)\big) = 0 .$$

Integration and assuming $\phi(\xi) \to 0$ for $\xi \to \pm\infty$ leads to

$$2c\Big(\gamma_n\phi(\xi) - \frac{1}{3}\xi\phi'(\xi)\Big) - \frac{\partial_\theta^4\mathbb{A}(0)}{24}\phi'''(\xi) + V'_{nl}\big(\phi'(\xi)\big) = 0 . \tag{3.7}$$

In case $\beta = 3$ with $V_{nl}(r) = \frac{1}{4}r^4$ we may simplify this equation. Due to $\gamma_3 = 0$ introducing

$$\zeta = -\frac{16c}{12^{1/3}(\partial_\theta^4\mathbb{A}(0))^{2/3}}\,\xi , \qquad \text{and} \qquad \psi(\zeta) = \Big(\frac{12}{\partial_\theta^4\mathbb{A}(0)}\Big)^{1/3}\phi'(\xi)$$

yields

$$\psi'' = 2\psi^3 + \zeta\psi \tag{3.8}$$

This is the second *Painlevé equation* P_{II}, cf. [Con99, IKSY91].

Now we consider (3.2). Again, due to $\omega(0) = 0$ the modulation term in (1.7) is obsolete. We again choose the ansatz

$$r_j(t) = R_j^\varepsilon(t) + \mathcal{O}(\varepsilon^{\gamma+1}) = \varepsilon^\gamma B\left(\varepsilon^3 t, \varepsilon(j - ct)\right) + \mathcal{O}(\varepsilon^{\gamma+1}) \tag{3.9}$$

for $j \in \mathbb{Z}$. For the linear terms holds

$$(\partial_t^2 + \mathcal{A})_j\mathbf{R}^\varepsilon = \left(-2c\,\partial_{\tau\eta}B + \frac{\partial_\theta^4\mathbb{A}(0)}{24}\,\partial_\eta^4 B\right)\,\varepsilon^{4+\gamma} + \mathcal{O}(\varepsilon^{5+\gamma}) .$$

Using Taylor expansion of the nonlinear term reads as

$$V'_{\mathrm{nl}}\big(R_{j+1}^\varepsilon\big) - 2V'_{\mathrm{nl}}\big(R_j^\varepsilon\big) + V'_{\mathrm{nl}}\big(R_{j-1}^\varepsilon\big) = \varepsilon^{\beta\gamma+2}\partial_\eta^2 V'_{\mathrm{nl}}(B) + \mathcal{O}(\varepsilon^{\beta\gamma+4}) .$$

Balancing linear and nonlinear terms leads to $\gamma = \frac{2}{\beta-1} =: \gamma_{\mathrm{r}}(\beta)$. Then employing in (3.2) gives the desired gKdV equation

$$2c\,\partial_{\tau\eta}B - \frac{\partial_\theta^4\mathbb{A}(0)}{24}\,\partial_\eta^4 B + \partial_\eta^2 V'_{\mathrm{nl}}(B) = 0 \tag{3.10}$$

for the macroscopic distance B. Now, the Ansatz

$$B(\tau, \eta) = \tau^{\tilde{\gamma}}\varphi\Big(\frac{\eta}{\tau^{1/3}}\Big) , \qquad \xi := \frac{\eta}{\tau^{1/3}} \tag{3.11}$$

yields

$$\partial_{\tau\eta} B = (\tilde{\gamma} - \frac{1}{3})\tau^{\tilde{\gamma}-4/3}\varphi' - \frac{1}{3}\tau^{\tilde{\gamma}-4/3}\xi\varphi'' , \quad \partial_\eta^4 B = \tau^{\tilde{\gamma}-4/3}\varphi^{(4)}$$

$$\partial_\eta^2 V_{\mathrm{nl}}'(B) = \tau^{\beta\tilde{\gamma}-(\beta+1)/3}\frac{\mathrm{d}^2}{\mathrm{d}\xi^2}V_{nl}'\big(\varphi(\xi)\big) .$$

Employing in (3.10) and using $\tilde{\gamma} = -\frac{1}{3}\gamma_{\mathrm{r}}(\beta) = -\frac{2}{3(\beta-1)} =: \tilde{\gamma}_{\mathrm{r}}(\beta)$ leads to

$$2c\Big((\tilde{\gamma} - \frac{1}{3})\varphi'(\xi) - \frac{1}{3}\xi\varphi''(\xi)\Big) - \frac{\partial_\theta^4\mathbb{A}(0)}{24}\varphi^{(4)}(\xi) + \frac{\mathrm{d}^2}{\mathrm{d}\xi^2}V_{\mathrm{nl}}'\big(\varphi(\xi)\big) = 0$$

Integration and assuming $\varphi(\xi) \to 0$ for $\xi \to +\infty$ finally leads to

$$2c\Big(\tilde{\gamma}\varphi - \frac{1}{3}\xi\varphi'(\xi)\Big) - \frac{\partial_\theta^4\mathbb{A}(0)}{24}\varphi'''(\xi) + \frac{\mathrm{d}}{\mathrm{d}\xi}V_{\mathrm{nl}}'\big(\varphi(\xi)\big) = 0 . \tag{3.12}$$

with $\tilde{\gamma} = \tilde{\gamma}_{\mathrm{r}}(\beta) := -\frac{2}{3(\beta-1)}$.

In case $\beta = 3$ this equation again simplifies. Due to $\tilde{\gamma}_{\mathrm{r}}(3) = -1/3$ we may integrate again. Assuming $\varphi(\xi) \to 0$ for $\xi \to +\infty$ we obtain

$$-\frac{2c}{3}\xi\varphi - \frac{\partial_\eta^4\mathbb{A}}{24}\varphi'' + V_{\mathrm{nl}}'(\varphi) = 0 \tag{3.13}$$

Thus, if $V_{\mathrm{nl}}(r) = \frac{1}{4}r^4$, then, apart from the constant coefficients which are to be eliminated by scaling the variables, we again obtain the second *Painlevé equation* P_{II}.

3.2 Simulation Results

Now we discuss some simulation results.

The first two plots, Figure 6 and 7, respectively, show linear behavior at the front for $\beta = 5$ and $\beta = 4$. The profiles match to the bounded solution of Airy's equation. The amplitude of Airy's function is chosen according to that of the linearized system, cf. (2.11).

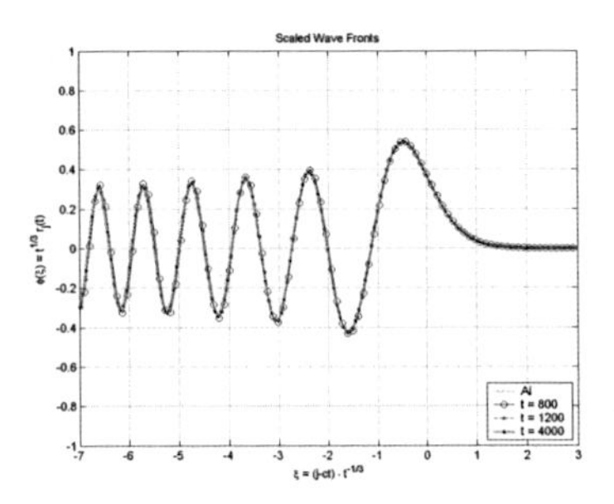

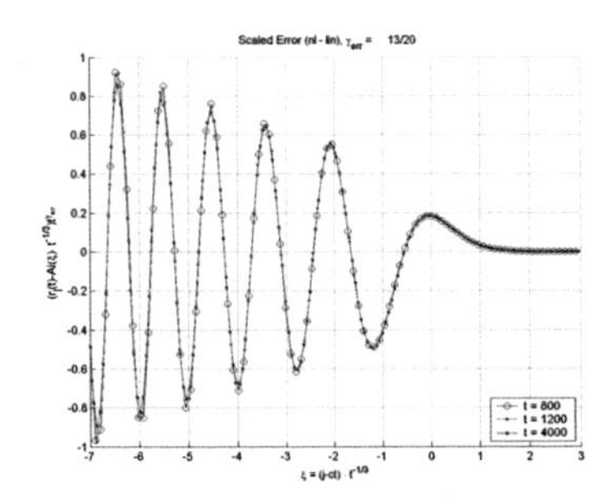

Figure 3.1: **Convergence to Airy-profile**: $\mathcal{N}(r) = r^5$, Riemann IC, amplitude scaling $\gamma = 1/3$

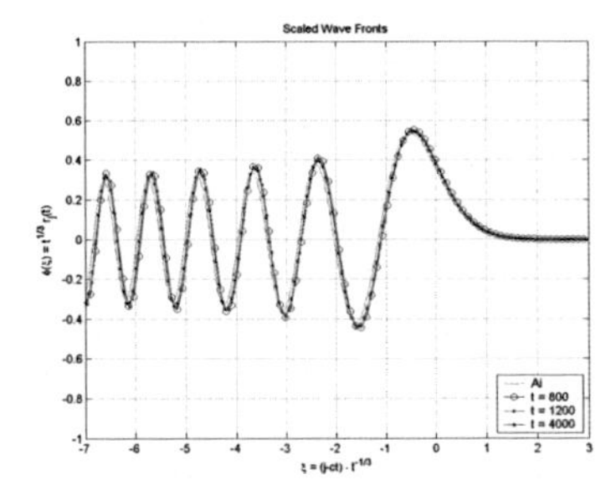

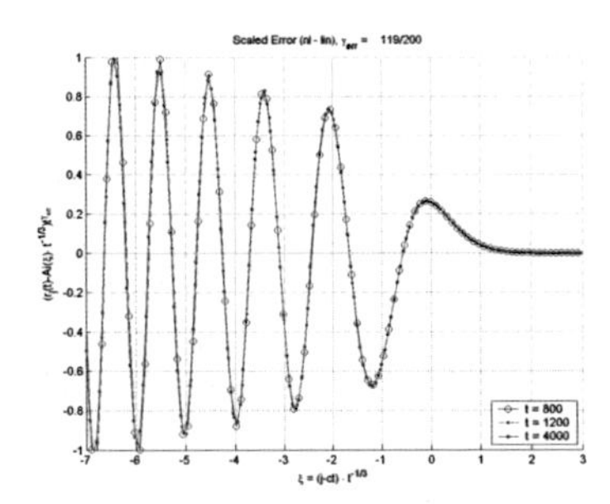

Figure 3.2: **Convergence to Airy-profile**: $\mathcal{N}(r) = r|r|^3$, Riemann IC, amplitude scaling $\gamma = 1/3$

Figure 8 shows the exact match to a solution of the Painlevé equation in case $\beta = 3$. To illustrate the difference the Airy function is also plotted.

In Figure 9 we may follow the passage from Airy- to the Painlevé profile as β varies in the region $(3, 4)$. Unfortunately ,in this case we do not even have a formal explanation.

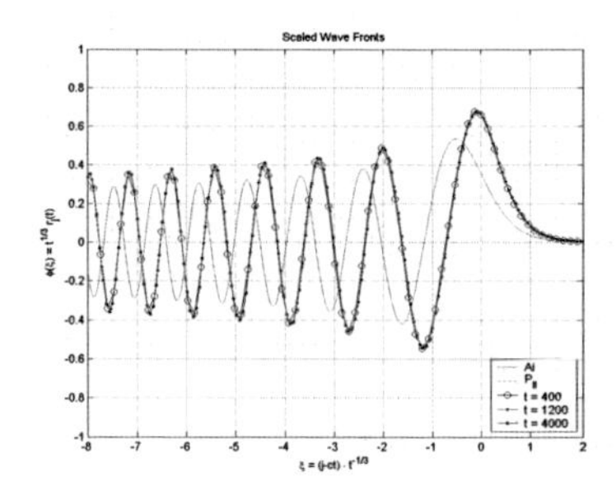

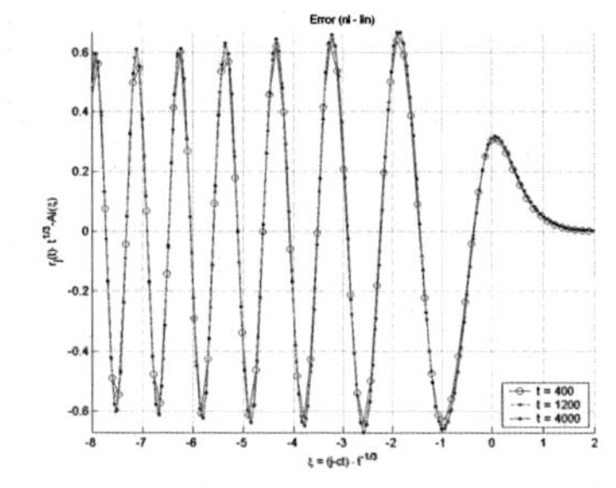

Figure 3.3: **Convergence to Painlevé-profile**: $V_{nl}(r) = 1/4r^4$, Riemann IC, amplitude scaling $\gamma = 1/3$

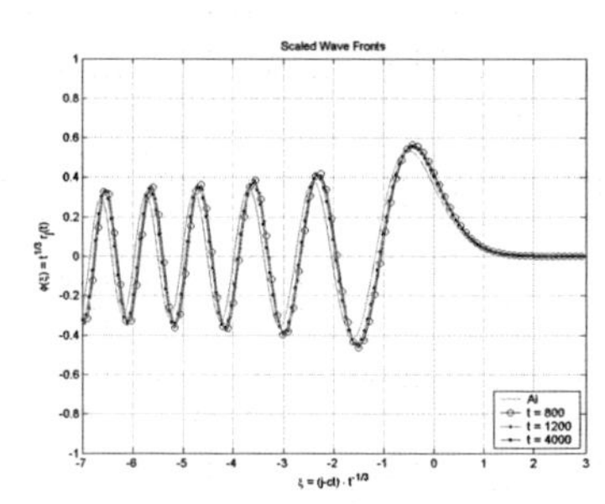

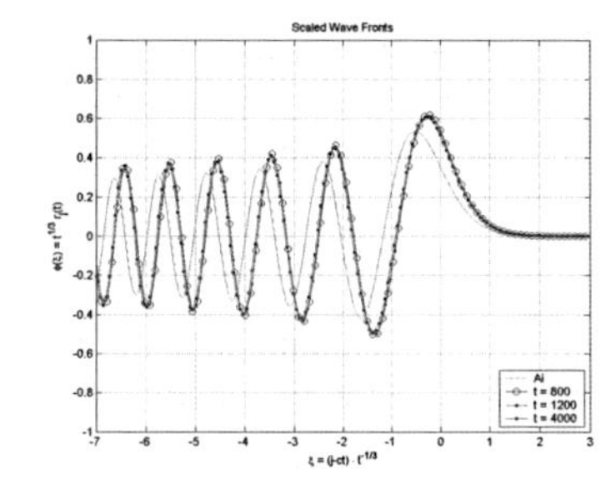

Figure 3.4: **Passage from linear to nonlinear behavior at the front**: Wave fronts for $\mathcal{N}(r) = r|r|^{2.7}$ (left) and $\mathcal{N}(r) = r|r|^{2.2}$ (right) to Riemann IC with, amplitude scaling $\gamma = 1/3$.

Finally, in Figure 10 and 11 we give an example of solitary waves occurring if the weight of the nonlinear term is increased.

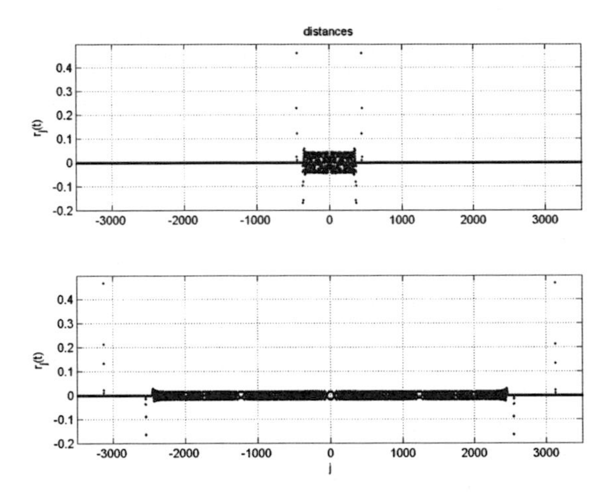

Figure 3.5: Solitary waves: $V_{nl}(r) = 5/4r^4$ with Riemann IC.

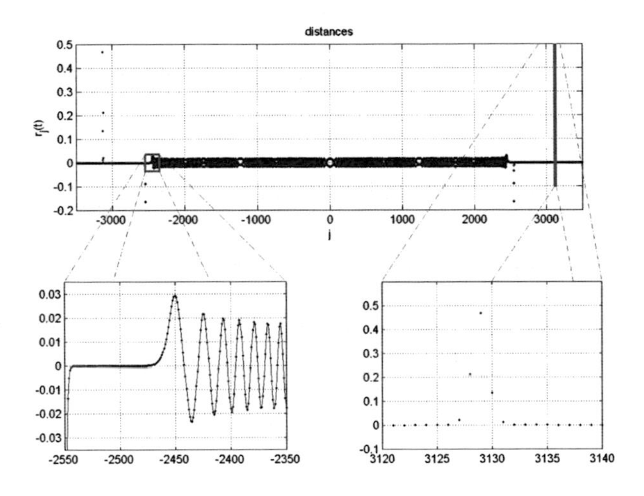

Figure 3.6: Solitary waves: $V_{nl}(r) = 5/4r^4$ with Riemann IC.

A The multi-scale approach

This section serves to provide some formulas to be used in the following. To keep to a simple notation we state everything in terms of $\mathbf{x}$, i.e. absolute displacements, although everything holds also true for the FPU lattice in terms of relative displacements.

A.1 KdV scaling and useful formulas

Expansion of the linear terms

For the linearized system of equations

$$\ddot{x}_j = -\mathcal{A}_j \mathbf{x}\,, \qquad j \in \mathbb{Z} \tag{A.1}$$

the scaling factor ε^γ for the amplitude is not relevant. Furthermore the complex conjugated terms do not interact. Thus it is sufficient to look at

$$X^\varepsilon_{n,j}(t) := A\left(\varepsilon^3 t, \varepsilon(j-ct)\right) \mathrm{e}^{\mathrm{i}n(\omega(\theta_0)t - j\theta_0)} = A(\tau,\eta) E^n_j(t)\,, \quad j \in \mathbb{Z} \tag{A.2}$$

for $n \in \mathbb{Z}$. Here we also skipped the indices of $A_{l,n}$ to shorten the notation.

Lemma A.1:

$$\mathrm{i}^m \partial^m_\theta \mathbb{A}(n\theta_0) = \sum_k a_k k^m \mathrm{e}^{-\mathrm{i}nk\theta_0} \tag{A.3}$$

Proof. The symbol is given by $\mathbb{A}(\theta) = \sum_k a_k \mathrm{e}^{\mathrm{i}k\theta}$, thus $\partial^m_\theta \mathbb{A}(\theta) = \mathrm{i}^m \sum_k a_k k^m \mathrm{e}^{\mathrm{i}k\theta}$. Since $a_k = a_{-k}$, i.e. $\mathbb{A}(\theta) = \mathbb{A}(-\theta)$, we obtain (A.3) □

Lemma A.2:
Consider the linear operator $\mathcal{A}$ defined by (1.4). Then for the ansatz(A.2) and $n \in \mathbb{Z}$ holds

$$\begin{aligned} \partial^2_t X^\varepsilon_{n,j}(t) = \Big(& -n^2\omega^2 A - 2\mathrm{i}n\omega c\, \partial_\eta A\, \varepsilon + c^2 \partial^2_\eta A\, \varepsilon^2 \\ & + 2\mathrm{i}n\omega\, \partial_\tau A\, \varepsilon^3 - 2c\, \partial_{\tau\eta} A\, \varepsilon^4 + \partial^2_\tau A\, \varepsilon^6 \Big) E^n_j \end{aligned} \tag{A.4}$$

where ω is evaluated in θ_0 and

$$\mathcal{A}_j \mathbf{X}^\varepsilon_n(t) = \sum_{m=0}^{M} \frac{\mathrm{i}^m \partial^m_\theta \mathbb{A}(n\theta_0)}{m!}\, \partial^m_\eta A(\tau,\eta)\, \varepsilon^m E^n_j + \mathcal{O}(\varepsilon^{M+1})\,. \tag{A.5}$$

Proof. Equation (A.4) is a simple consequence of $\partial_t A = \varepsilon^3 \partial_\tau A - \varepsilon c\, \partial_\eta A$ and $\partial_t E^n_j = \mathrm{i}n\omega E^n_j$.

To proof formula (A.5) note that for the ansatz (A.2) holds

$$X^{\varepsilon}_{n,j+k}(t) = \sum_{m=0}^{M} \frac{1}{m!}\,\partial_\eta^m A(\tau,\eta)\,(k\varepsilon)^m \mathrm{e}^{-\mathrm{i}nk\theta} E_j^n + \mathcal{O}(\varepsilon^{M+1})\,.$$

Together with (A.3) we get

$$\begin{aligned}
\mathcal{A}_j\,\mathbf{X}^{\varepsilon}_n(t) &= \sum_k a_k \left(\sum_{m=0}^{M} \frac{1}{m!}\,\partial_\eta^m A(\tau,\eta)\,(k\varepsilon)^m\right) \mathrm{e}^{-\mathrm{i}nk\theta_0} E_j^n + \mathcal{O}(\varepsilon^{M+1}) \\
&= \sum_{m=0}^{M} \frac{1}{m!}\,\partial_\eta^m A(\tau,\eta) \left(\sum_k a_k k^m \mathrm{e}^{-\mathrm{i}nk\theta_0}\right) \varepsilon^m E_j^n + \mathcal{O}(\varepsilon^{M+1}) \\
&= \sum_{m=0}^{M} \frac{i^m \partial_\theta^m \mathbb{A}(n\theta_0)}{m!}\,\partial_\eta^m A(\tau,\eta)\,\varepsilon^m E_j^n + \mathcal{O}(\varepsilon^{M+1})\,.
\end{aligned}$$

□

Formula (A.5) will be of particular interest for $n=0$ (or $\theta_0=0$) and $|n|=1$. In case $n=0$ the right hand side simplifies due to the symmetry $\mathbb{A}(\theta)=\mathbb{A}(-\theta)$. If ω is smooth in θ_0 and θ_0 corresponds to some wave front then we also may rewrite the right hand side of (A.5). Differentiating the dispersion relation $\mathbb{A}(\theta)=\omega^2(\theta)$ and keeping in mind that the wave front is characterized by $\omega''(\theta_0)=0$ we obtain the second part of the following corollary.

Corollary A.3:
Consider the linear operator $\mathcal{A}$ and $\mathbf{X}^{\varepsilon}_n$ defined by (1.4) and (A.2), respectively.

1. *For $n=0$ or $\theta_0=0$ holds*

$$\mathcal{A}_j\,\mathbf{X}^{\varepsilon}_n = \Big(\mathbb{A}(0)A - \frac{\partial_\theta^2\mathbb{A}(0)}{2}\,\partial_\eta^2 A\,\varepsilon^2 + \frac{\partial_\theta^4\mathbb{A}(0)}{24}\,\partial_\eta^4 A\,\varepsilon^4\Big) E_j^n + \mathcal{O}(\varepsilon^6)\,. \qquad \text{(A.6)}$$

2. *Assume ω to be smooth and $\omega''(\theta_0)=0$. Then*

$$\begin{aligned}
\mathcal{A}_j\,\mathbf{X}^{\varepsilon}_{\pm1} = \Big(&\omega^2 A \pm 2\mathrm{i}\omega\omega'\,\partial_\eta A\,\varepsilon - (\omega')^2\partial_\eta^2 A\,\varepsilon^2 \mp \frac{\mathrm{i}}{3}\omega\omega'''\partial_\eta^3 A\,\varepsilon^3 \\
&+ \frac{1}{12}(4\omega'\omega''' + \omega\omega^{(4)})\partial_\eta^4 A\,\varepsilon^4\Big) E_j^{\pm1} + \mathcal{O}(\varepsilon^5)\,,
\end{aligned} \qquad \text{(A.7)}$$

where ω, ω' etc. are evaluated in $\theta=\theta_0$.

Reduction of the symplectic structure

Inserting ansatz functions of the form (1.7) into the infinite dimensional ODE we will derivate macroscopic PDE's for the modulated patterns A_n. A more natural second way to derivate these equations is the reduction of Hamiltonian or Lagrangian structures. By doing so the underlying geometry and physics becomes much more clear.

The macroscopic Lagrangian and Hamiltonian functions will be deviated in the upcoming sections. Here we only want to provide the symplectic structure for later use.

The canonical symplectic structure on $l^2(\mathbb{Z})$, or more precisely on the cotangent bundle $T^*l^2(\mathbb{Z})$ is $\Omega((\mathbf{x},\mathbf{p}),(\tilde{\mathbf{x}},\tilde{\mathbf{p}})) = \sum_{j\in\mathbb{Z}} x_j\tilde{p}_j - \tilde{x}_j p_j$. For $(\mathbf{x},\mathbf{p})$ the ansatz (1.7) induces

$$(\mathbf{x},\mathbf{p}) = R_\varepsilon(A) = (\varepsilon A, \varepsilon^4\partial_\tau A - c\varepsilon^2\partial_\eta A)\ .$$

We embed the discrete system in $L^2(\mathbb{R})$ by piecewise linear interpolation. Then the canonical symplectic structure transforms as follows

$$\begin{aligned}\Omega((\mathbf{x},\mathbf{p}),(\tilde{\mathbf{x}},\tilde{\mathbf{p}})) &= \int \Big(A(\varepsilon^4\partial_\tau\tilde{A} - c\varepsilon^2\partial_\eta\tilde{A}) - \tilde{A}(\varepsilon^4\partial_\tau A - c\varepsilon^2\partial_\eta A)\Big)\,\mathrm{d}\eta\\ &= \varepsilon^2 c\int\Big(\tilde{A}\partial_\eta A - A\partial_\eta\tilde{A}\Big)\,\mathrm{d}\eta + \mathcal{O}(\varepsilon^4)\\ &= \varepsilon^2\Omega^{\mathrm{red}}(A,\tilde{A}) + \mathcal{O}(\varepsilon^4)\end{aligned}$$

for

$$\Omega^{\mathrm{red}}(A,\tilde{A}) = 2c\int \partial_\eta A\,\tilde{A}\,\mathrm{d}\eta\ .$$

The linear map $\mathbb{S}^{\mathrm{red}} : TL^2(\mathbb{R}) \to T^*L^2(\mathbb{R})$ associated with the reduced symplectic form Ω^{red} reads as

$$\mathbb{S}^{\mathrm{red}} = 2c\partial_\eta \qquad (A.8)$$

Then the reduced Hamiltonian system is given by

$$\mathbb{S}^{\mathrm{red}}\partial_\tau A = \mathrm{d}\,\mathcal{H}^{\mathrm{red}}(A)\ . \qquad (A.9)$$

A.2 Linear and nonlinear consistency, self-similiar solutions

To end this section we want to give a brief motivation for using the KdV scaling (1.6). The choice of the spacial variable η is obvious. For the time variable we may assume a more general scaling factor $\tau := \varepsilon^\alpha t$.

Let us assume for simplicity $\theta_0 = 0$ and $\mathbb{A}(0) = 0$, i.e. in particular we have no modulation term. At first we consider the linear terms. Using the ansatz $X_j^\varepsilon(t) = \varepsilon^\gamma A(\varepsilon^\alpha t, \varepsilon(j-ct))$ and keeping in mind that $2c^2 = \partial_\theta^2\mathbb{A}(0)$ equation (2.1) turns into

$$-2c\partial_{\tau\eta}A\,\varepsilon^{\alpha+\gamma+1} + \frac{\partial_\theta^4\mathbb{A}(0)}{24}\,\partial_\eta^4 A\,\varepsilon^{4+\gamma} = \mathcal{O}(\varepsilon^{5+\gamma})\ .$$

The aim is to derivate a PDE for A which describes the wave fronts of the discrete system. Obviously we obtain a non-trivial PDE that admits an solution which at least qualitatively describes the the wave front if and only if $\alpha = 3$. We refer to this as *linear consistency* condition.

The scaling factor ε^γ for the amplitudes depends on the nonlinear term. To determine γ we assume that employing the ansatz X_j^ε in the nonlinear terms leads to

$$\mathcal{N}_j(\mathbf{X}_j^\varepsilon) = \varepsilon^{\beta\gamma+\beta_0}\,\mathcal{F}(A) + o(\varepsilon^{\beta\gamma+\beta_0})$$

where $\mathcal{F}$ is a nonlinear operator acting on $A(\tau,\cdot)$ and $\beta, \beta_0 \in \mathbb{R}$. Combining the linear and nonlinear terms we end up with

$$2c\partial_{\tau\eta}A\,\varepsilon^{\alpha+\gamma+1} - \frac{\partial_\theta^4\mathbb{A}(0)}{24}\,\partial_\eta^4 A\,\varepsilon^{4+\gamma} + \varepsilon^{\beta\gamma+\beta_0}\,\mathcal{F}(A) = o(\varepsilon^{4+\gamma} + \varepsilon^{a(\gamma)})\,.$$

Again we aim to derive a non-trivial PDE for the wave fronts. Balancing the linear and nonlinear interaction implies $4+\gamma = \beta\gamma+\beta_0$ to which we refer to as *nonlinear consistency*.

The scaling exponents α and γ carry over to the ansatz for the self-similar solution of the PDE. To see this note that we have the scaling property

$$\mathcal{F}\left(\varepsilon^\gamma A(\tau,\varepsilon\cdot)\right) = \varepsilon^{\beta\gamma+\beta_0}\,\mathcal{F}\left(A(\tau,\cdot)\right)\,.$$

Thus the PDE $\partial_{\tau\eta}A + C\partial_\eta^4 A + \mathcal{F}(A) = 0$ admits a solution of the form

$$A(\tau,\eta) = \tau^{\tilde{\gamma}}\phi(\tau^{\tilde{\alpha}}\eta)$$

if and only if

$$\tilde{\alpha} = -\frac{1}{\alpha} = -\frac{1}{3}\,, \qquad \text{and} \qquad \tilde{\gamma} = \tilde{\alpha}\gamma = -\frac{\gamma}{\alpha}\,.$$

B Nonlinear KG-type systems

Now we consider systems of particles interacting with quadratic pair potentials, i.e. linear forces, but nonlinear background potentials. The simplest example is the *Klein-Gordon chain* (KG)

$$\ddot{x}_j = x_{j-1} - 2x_j + x_{j+1} - W'(x_j)\,.$$

Like in the FPU case in the previous section we allow for a more general linear interaction. Thus we consider the generalized KG system

$$\ddot{x}_j = \mathcal{A}_j\,\mathbf{x} - W'_{nl}(x_j)\,. \tag{B.1}$$

where $\mathcal{A}$ is a linear operator of the form (1.4) and W'_{nl} refers to the nonlinear part of the on-site potential. We assume W_{nl} to be sufficiently smooth and $W'_{nl}(0) = W''_{nl}(0) = 0$.

We choose the ansatz (1.7) for $L = 3$. That is we we have

$$x_j(t) = X_j^\varepsilon(t) + o(\varepsilon^{\gamma_3}) \tag{B.2}$$

where

$$X_j^\varepsilon(t) = \sum_{l=1}^{3} \varepsilon^{\gamma_l} \sum_{n=-l}^{l} A_{l,n}\left(\varepsilon^3 t, \varepsilon(j-ct)\right) E^n(t) = \sum_{l=1}^{3}\sum_{n=-l}^{l} X_{l,n,j}^\varepsilon(t)$$

with $\gamma_1 < \gamma_2 < \gamma_3$ and $A_{l,n} = \bar{A}_{l,-n}$.

Linear terms

Depending on the order of the harmonic term E^n we obtain the following expansions for the linear terms. Without modulation term, i.e. in case $n = 0$, we use (A.6) to get

$$(\partial_t^2 - \mathcal{A})_j \mathbf{X}_{l,0}^\varepsilon = -\omega^2(0) A_{l,0}\varepsilon^{\gamma_l} + \left(c^2 - (\omega'(0))^2\right) \partial_\eta^2 A_{l,0}\varepsilon^{\gamma_l+2} + \mathcal{O}(\varepsilon^{\gamma_l+3})\,. \quad \text{(B.3)}$$

The term of order ε^{γ_l+2} does not vanish since in general $c^2 = (\omega'(\theta_0))^2 \neq (\omega'(0))^2$. In case of first order harmonics E^1 (A.7) implies

$$\begin{aligned}&(\partial_t^2 - \mathcal{A})_j \mathbf{X}_{l,1}^\varepsilon \\ &= \left(2\mathrm{i}\omega(\theta_0)\partial_\tau A_{l,1} - \frac{\mathrm{i}}{3}\omega(\theta_0)\omega'''(\theta_0)\partial_\eta^3 A_{l,1}\right) E\,\varepsilon^{\gamma_l+3} + \text{c.c.} + \mathcal{O}(\varepsilon^{\gamma_l+4})\end{aligned} \quad \text{(B.4)}$$

For terms with second and third order harmonics E^2 and E^3, respectively, we obtain from (A.5)

$$(\partial_t^2 - \mathcal{A})_j \mathbf{X}_{l,2}^\varepsilon = \big(-4\omega(\theta_0) + \omega(2\theta_0)\big) A_{l,2} E^2\,\varepsilon^{\gamma_l} + \text{c.c.} + \mathcal{O}(\varepsilon^{\gamma_l+1}) \quad \text{(B.5)}$$

$$(\partial_t^2 - \mathcal{A})_j \mathbf{X}_{l,3}^\varepsilon = \big(-9\omega(\theta_0) + \omega(3\theta_0)\big) A_{l,3} E^3\,\varepsilon^{\gamma_l} + \text{c.c.} + \mathcal{O}(\varepsilon^{\gamma_l+1})\,. \quad \text{(B.6)}$$

Nonlinear terms

$$\begin{aligned}(X_j^\varepsilon)^2 =& \Big(A_{1,0}^2 + 2|A_{1,1}|^2 + 2A_{1,0}A_{1,1}E + A_{1,1}^2E^2\Big)\varepsilon^{2\gamma_1} \\ &+ 2\Big(A_{1,0}A_{2,0} + \bar{A}_{1,1}A_{2,1} + (A_{1,0}A_{2,1} + A_{1,1}A_{2,0} + \bar{A}_{1,1}A_{2,2})E \\ &\qquad + (A_{1,0}A_{2,2} + A_{1,1}A_{2,1})E^2 + A_{1,1}A_{2,2}E^3\Big)\varepsilon^{\gamma_1+\gamma_2} \\ &+ \Big(A_{2,0}^2 + 2|A_{2,1}|^2 + 2|A_{2,2}|^2 + 2(A_{2,0}A_{2,1} + \bar{A}_{2,1}A_{2,2})E \\ &\qquad + (A_{2,1}^2 + 2A_{2,0}A_{2,2})E^2 + 2A_{2,1}A_{2,2}E^3 + A_{2,2}E^4\Big)\varepsilon^{2\gamma_2} \\ &+ \text{c.c.} + \mathcal{O}(\varepsilon^{\gamma_1+\gamma_3})\end{aligned} \quad \text{(B.7)}$$

$$\begin{aligned}(X_j^\varepsilon)^3 =& \Big(A_{1,0}^3 + 6A_{1,0}|A_{1,1}|^2 + 3(A_{1,0}^2A_{1,1} + |A_{1,1}|^2A_{1,1})E \\ &\qquad + 3A_{1,0}A_{1,1}^2E^2 + A_{1,1}^3E^3\Big)\varepsilon^{3\gamma_1} + \text{c.c.} + \mathcal{O}(\varepsilon^{2\gamma_1+\gamma_2})\end{aligned} \quad \text{(B.8)}$$

Modulation equation (quadratic nonlinearities)

We rewrite (B.1) using Taylor expansion for the nonlinear force on the right hand side. Keeping in mind that $W''_{nl}(0) = 0$ we obtain

$$(\partial_t^2 - \mathcal{A})_j \mathbf{X}^\varepsilon = -\frac{W_{nl}^{(3)}(0)}{2}(X_j^\varepsilon)^2 - \frac{W_{nl}^{(4)}(0)}{6}(X_j^\varepsilon)^3 + \mathcal{O}(\dots)\,. \tag{B.9}$$

The modulation equations are obtained using (B.3) - (B.8) and equating coefficients of the terms $\varepsilon^{\gamma_l} E^n$. It is sufficient to consider $n \geq 0$ since the other terms are just the complex conjugates.

The lowest order term is

$$\varepsilon^{\gamma 1}: \qquad -\omega^2(0) A_{1,0} = 0$$

i.e. $A_{1,0} = 0$. Thus (B.7) and (B.8) simplify substantially. We proceed with imposing conditions on γ_1 and γ_2. We aim to derive an equation for $A_{1,1}$, in particular we have $A_{1,1} \neq 0$. The leading order term corresponding to $A_{1,1}$ on the left hand side is of order $\varepsilon^{\gamma_1+3}E$. The lowest order term of the first harmonic on the right hand side is contains the factor $\varepsilon^{\gamma_1+\gamma_2}E$. To obtain an equation effected by the nonlinear term we have to require $\gamma_1 + 3 = \gamma_1 + \gamma_2$, i.e. $\gamma_2 = 3$. It remains $2|A_{1,1}|^2\varepsilon^{2\gamma_1}$ on the right hand side. To provide a counterpart for this term on the left hand side we impose the condition $\gamma_2 = 2\gamma_1$, i.e. $\gamma_1 = \frac{3}{2}$. Then

$$\begin{aligned} \varepsilon^3: \qquad -\omega^2(0)A_{2,0} &= -\frac{W_{nl}^{(3)}(0)}{2} \cdot 2|A_{1,1}|^2 \\ \Longleftrightarrow \qquad A_{2,0} &= \frac{W_{nl}^{(3)}(0)}{\omega^2(0)}\,|A_{1,1}|^2 \end{aligned} \tag{B.10}$$

and $A_{2,0}$ is determined by $A_{1,1}$ which remains free at this stage.

To summarize, we have

$$\gamma_1 = \frac{3}{2}\,, \qquad \gamma_2 = 3$$

and $A_{1,0} = 0$. Thus the left hand side of (B.9) reads as

$$\begin{aligned} (\partial_t^2 - \mathcal{A})_j \mathbf{X}^\varepsilon = &\Big(- \omega(0)A_{2,0} + \left(-4\omega^2(\theta_0) + \omega^2(2\theta_0)\right) A_{2,2}E^2\Big)\varepsilon^3 \\ &+ \Big(2\mathrm{i}\omega(\theta_0)\partial_\tau A_{1,1} - \frac{\mathrm{i}}{3}\omega(\theta_0)\omega'''(\theta_0)\partial_\eta^3 A_{1,1}\Big) E\,\varepsilon^{\frac{9}{2}} \\ &+ \Big(- \omega(0)A_{3,0} + \left(-4\omega^2(\theta_0) + \omega^2(2\theta_0)\right) A_{3,2}E^2 \\ &\quad + \left(-9\omega^2(\theta_0) + \omega^2(3\theta_0)\right) A_{3,3}E^3\Big)\varepsilon^{\gamma_3} + \text{c.c.} + \mathcal{O}(\varepsilon^5)\,. \end{aligned}$$

For the right hand side we obtain

$$
\begin{aligned}
&-\frac{W_{nl}^{(3)}(0)}{2}(X_j^\varepsilon)^2-\frac{W_{nl}^{(4)}(0)}{6}(X_j^\varepsilon)^3=-W_{nl}^{(3)}(0)\Big(|A_{1,1}|^2+\frac{1}{2}A_{1,1}^2E^2\Big)\varepsilon^3\\
&-\Big(\bar{A}_{1,1}A_{2,1}W_{nl}^{(3)}(0)+\Big((A_{1,1}A_{2,0}+\bar{A}_{1,1}A_{2,2})W_{nl}^{(3)}(0)+\frac{W_{nl}^{(4)}(0)}{2}|A_{1,1}|^2A_{1,1}\Big)E\\
&\quad+W_{nl}^{(3)}(0)A_{1,1}A_{2,1}E^2+\Big(W_{nl}^{(3)}(0)A_{1,1}A_{2,2}+\frac{W_{nl}^{(4)}(0)}{6}A_{1,1}^3\Big)E^3\Big)\varepsilon^{\frac{9}{2}}\\
&+\text{c.c.}+\mathcal{O}(\varepsilon^6)
\end{aligned}
$$

Now we continue to equating the coefficients in (B.9). The terms corresponding to ε^3E^0 are already exploited in (B.10). There is no term of order ε^3E. For the next harmonic we obtain

$$
\begin{aligned}
\varepsilon^3E^2: \qquad & \left(-4\omega^2(\theta_0)+\omega^2(2\theta_0)\right)A_{2,2}=-\frac{W_{nl}^{(3)}(0)}{2}A_{1,1}^2\\
\Longleftrightarrow \qquad & A_{2,2}=\frac{W_{nl}^{(3)}(0)}{8\omega^2(\theta_0)-2\omega^2(2\theta_0)}A_{1,1}^2
\end{aligned}
\tag{B.11}
$$

if and only if the *non-resonance condition*

$$
-\left(2\omega(\theta_0)\right)^2+\omega^2(2\theta_0)\neq 0 \tag{B.12}
$$

holds. This is necessary due to $A_{1,1}\neq 0$. Thus $A_{2,2}$ is also determined by $A_{1,1}$. In the latter case $A_{3,0}$ and $A_{2,1}$ are dependent.

Concerning terms of order $\varepsilon^{\frac{9}{2}}$ we find that either $A_{2,1}=0$ or $\gamma_3=\frac{9}{2}$ and

$$
\varepsilon^{\frac{9}{2}}: \qquad -\omega^2(0)A_{3,0}=-W_{nl}^{(3)}(0)\left(\bar{A}_{1,1}A_{2,1}+A_{1,1}\bar{A}_{2,1}\right)
$$

The next terms contribute to the PDE for $A_{1,1}$,

$$
\begin{aligned}
\varepsilon^{\frac{9}{2}}E^1: \qquad & 2\mathrm{i}\omega(\theta_0)\partial_\tau A_{1,1}-\frac{\mathrm{i}}{3}\omega(\theta_0)\omega'''(\theta_0)\partial_\eta^3A_{1,1}=\\
& -W_{nl}^{(3)}(0)\left(A_{1,1}A_{2,0}+\bar{A}_{1,1}A_{2,2}\right)-\frac{W_{nl}^{(4)}(0)}{2}|A_{1,1}|^2A_{1,1}\,.
\end{aligned}
$$

Using (B.10) and (B.11) we obtain the desired modulation equation

$$
\begin{aligned}
&2\mathrm{i}\omega(\theta_0)\partial_\tau A_{1,1}-\frac{\mathrm{i}}{3}\omega(\theta_0)\omega'''(\theta_0)\partial_\eta^3A_{1,1}=\\
&\left(-\frac{\left(W_{nl}^{(3)}(0)\right)^2}{\omega^2(0)}-\frac{\left(W_{nl}^{(3)}(0)\right)^2}{8\omega^2(\theta_0)-2\omega^2(2\theta_0)}-\frac{W_{nl}^{(4)}(0)}{2}\right)|A_{1,1}|^2A_{1,1}\,.
\end{aligned}
\tag{B.13}
$$

It remains to check the higher harmonics for consistency. Concerning terms of order $\varepsilon^{\frac{9}{2}}E^2$ we find that either $A_{2,1}=0$ or $\gamma_3=\frac{9}{2}$ and

$$
\varepsilon^{\frac{9}{2}}E^2: \qquad \left(-4\omega^2(\theta_0)+\omega^2(2\theta_0)\right)A_{3,2}=-W_{nl}^{(3)}(0)A_{1,1}A_{2,1}
$$

The last equation is solvable if the non-resonance condition (B.12) is satisfied. Terms corresponding to $\varepsilon^{\frac{9}{2}}$ remain. Looking first at the right hans side we find that either $W_{nl}^{(3)}(0)A_{1,1}A_{2,2} + \frac{W_{nl}^{(4)}(0)}{6}A_{1,1}^3 = 0$, i.e. using (B.11)

$$\frac{\left(W_{nl}^{(3)}(0)\right)^2}{8\omega^2(\theta_0) - 2\omega^2(2\theta_0)} - \frac{W_{nl}^{(4)}(0)}{6} = 0$$

or we have to assume $\gamma_3 = \frac{9}{2}$ and get

$$\varepsilon^3 E^3: \qquad \left(-9\omega^2(\theta_0) + \omega^2(3\theta_0)\right) A_{2,2} = -W_{nl}^{(3)}(0)A_{1,1}A_{2,2} - \frac{W_{nl}^{(4)}(0)}{6} A_{1,1}^3 \,.$$

The last equation is solvable if and only if the *non-resonance condition*

$$-\left(3\omega(\theta_0)\right)^2 + \omega^2(3\theta_0) \neq 0 \tag{B.14}$$

is satisfied. Aiming to the most general case we assume that the latter holds. In that case $A_{3,3}$ can be expressed in terms of $A_{1,1}$.

Note finally that we did not get an equation depending on $A_{3,1}$. It only contributes to higher order terms. I.e. we may choose $A_{3,1} = 0$. Furthermore, although we have chosen $\gamma_3 = \frac{9}{2}$ the set of equations remains consistent if we also set $A_{2,1}, A_{3,0}, A_{3,2} = 0$. Thus the ansatz

$$X_j^\varepsilon(t) = \varepsilon^{\frac{3}{2}} A_{1,1}E + \varepsilon^3\left(A_{2,0} + A_{2,2}E^2\right) + \varepsilon^{\frac{9}{2}} A_{3,3}E^3$$

would have been sufficient.

Proposition B.1:
Assume that the Klein-Gordon type system (B.1) *admits solutions of the form*

$$x_j(t) = \sum_{l=1}^{3} \varepsilon^{\gamma_l} \sum_{n=-l}^{l} A_{l,n}\left(\varepsilon^3 t, \varepsilon(j-ct)\right) E^n(t) + o(\varepsilon^{\gamma_3})$$

where $\gamma_1 < \gamma_2 < \gamma_3$ *for all* $\varepsilon \in (0, \varepsilon_0)$ *for some* $\varepsilon_0 > 0$ *and that the non-resonance conditions* (B.12) *and* (B.14) *are satisfied. Then* $A_{1,0} = 0$ *and* $A_{1,1}$ *satisfies the PDE*

$$\partial_\tau A + C_1 \partial_\eta^3 A = \mathrm{i} C_2 |A|^2 A \tag{B.15}$$

where

$$C_1 = -\frac{\omega'''(\theta_0)}{6}$$

$$C_2 = \frac{\left(W_{nl}^{(3)}(0)\right)^2}{\omega^2(0)} + \frac{\left(W_{nl}^{(3)}(0)\right)^2}{8\omega^2(\theta_0) - 2\omega^2(2\theta_0)} + \frac{W_{nl}^{(4)}(0)}{2} \,.$$

References

[BH86] N. BLEISTEIN and R. A. HANDELSMAN. *Asymptotic expansions of integrals.* Dover Publications Inc., New York, second edition, 1986.

[Con99] R. Conte, editor. *The Painlevé property.* CRM Series in Mathematical Physics. Springer-Verlag, New York, 1999. One century later.

[FM02] G. FRIESECKE and K. MATTHIES. Atomic-scale localization of high-energy solitary waves on lattices. *Phys. D*, 171(4), 211–220, 2002.

[FP99] G. FRIESECKE and R. L. PEGO. Solitary waves on FPU lattices. I. Qualitative properties, renormalization and continuum limit. *Nonlinearity*, 12(6), 1601–1627, 1999.

[FP02] G. FRIESECKE and R. L. PEGO. Solitary waves on FPU lattices. II. Linear implies nonlinear stability. *Nonlinearity*, 15(4), 1343–1359, 2002.

[FP04a] G. FRIESECKE and R. L. PEGO. Solitary waves on Fermi-Pasta-Ulam lattices. III. Howland-type Floquet theory. *Nonlinearity*, 17(1), 207–227, 2004.

[FP04b] G. FRIESECKE and R. L. PEGO. Solitary waves on Fermi-Pasta-Ulam lattices. IV. Proof of stability at low energy. *Nonlinearity*, 17(1), 229–251, 2004.

[FPU55] E. FERMI, J. PASTA, and S. ULAM. Studies of nonlinear problems. *Los Alamos Scientific Laboratory of the University of California*, Report LA-1940, 1955.

[Fri03] G. FRIESECKE. Dynamics of the infinite harmonic chain: conversion of coherent initial data into synchronized binary oscillations. *Preprint*, 2003.

[FW94] G. FRIESECKE and J. A. D. WATTIS. Existence theorem for solitary waves on lattices. *Comm. Math. Phys.*, 161(2), 391–418, 1994.

[GHM06] J. GIANNOULIS, M. HERRMANN, and A. MIELKE. Continuum descriptions for the dynamics in discrete lattices: derivation and justification. In *Analysis, modeling and simulation of multiscale problems*, pages 435–466. Springer, Berlin, 2006.

[GM06] J. GIANNOULIS and A. MIELKE. Dispersive evolution of pulses in oscillator chains with general interaction potentials. *Discrete Contin. Dyn. Syst. Ser. B*, 6(3), 493–523 (electronic), 2006.

[Hör90] L. HÖRMANDER. *The analysis of linear partial differential operators. I*, volume 256 of *Grundlehren der Mathematischen Wissenschaften.* Springer-Verlag, Berlin, second edition, 1990.

[IKSY91] K. Iwasaki, H. Kimura, S. Shimomura, and M. Yoshida. *From Gauss to Painlevé.* Aspects of Mathematics, E16. Friedr. Vieweg & Sohn, Braunschweig, 1991. A modern theory of special functions.

[McM02] E. McMillan. Multiscale correction to solitary wave solutions on FPU lattices. *Nonlinearity*, 15(5), 1685–1697, 2002.

[Mie06] A. Mielke. Macroscopic behavior of microscopic oscillations in harmonic lattices via Wigner-Husimi transforms. *Arch. Ration. Mech. Anal.*, 181(3), 401–448, 2006.

[MP10] A. Mielke and C. Patz. Dispersive stability of infinite-dimansional Hamiltonian systems on lattices. *Applicable Analysis*, 89(9), 1493–1512, 2010.

[MP13] A. Mielke and C. Patz. Uniform asymptotic expansions for the infinite harmonic chain. *WIAS Preprint*, 1846, 2013.

[RH93] P. Rosenau and J. M. Hyman. Compactons: Solitons with finite wavelength. *Phys. Rev. Lett.*, 70(5), 564–567, Feb 1993.

[SW00] G. Schneider and C. E. Wayne. Counter-propagating waves on fluid surfaces and the continuum limit of the Fermi-Pasta-Ulam model. In B. Fiedler, K. Gröger, and J. Sprekels, editors, *International Conference on Differential Equations*, volume 1, pages 390–404. World Scientific, 2000.

[Won89] R. Wong. *Asymptotic approximations of integrals.* Computer Science and Scientific Computing. Academic Press Inc., Boston, MA, 1989.

[ZK65] N. J. Zabusky and M. D. Kruskal. Interaction of "solitons" in a collisionless plasma and the recurrence of initial states. *Phys. Rev. Lett.*, 15(6), 240–243, Aug 1965.

Chapter 5

Finite-temperature coarse-graining of one-dimensional models: Mathematical analysis and computational approaches

A earlier version following paper by X. BLANC, C. LE BRIS, F. LEGOLL and C. PATZ is available as *INRIA Rapport de Recherche No. 6544, May 2008.* It was finally published in *Journal of Nonlinear Science, Volume 20, Number 2, 2010, p. 241-275.*

The authors present a possible approach for the computation of free energies and ensemble averages of one-dimensional coarse-grained models in materials science. The approach is based upon a thermodynamic limit process, and makes use of ergodic theorems and large deviations theory. In addition to providing a possible efficient computational strategy for ensemble averages, the approach allows for assessing the accuracy of approximations commonly used in practice.

Finite-temperature coarse-graining of one-dimensional models: mathematical analysis and computational approaches

X. Blanc[1,4], C. Le Bris[2,4], F. Legoll[3,4] and C. Patz [5]

[1] Laboratoire J.-L. Lions, Université Pierre et Marie Curie, Boîte courrier 187, 75252 Paris Cedex 05, France

[2] CERMICS, École Nationale des Ponts et Chaussées, Université Paris-Est, 6 et 8 avenue Blaise Pascal, 77455 Marne-La-Vallée Cedex 2, France

[3] Institut Navier, LAMI, École Nationale des Ponts et Chaussées, Université Paris-Est, 6 et 8 avenue Blaise Pascal, 77455 Marne-La-Vallée Cedex 2, France

[4] INRIA Rocquencourt, MICMAC team-project, Domaine de Voluceau, B.P. 105, 78153 Le Chesnay Cedex, France

[5] Weierstrass-Institut für Angewandte Analysis und Stochastik, Mohrenstrasse 39, 10117 Berlin, Germany

23 May 2008. Revised 9 November 2009

Abstract

We present a possible approach for the computation of free energies and ensemble averages of one-dimensional coarse-grained models in materials science. The approach is based upon a thermodynamic limit process, and makes use of ergodic theorems and large deviations theory. In addition to providing a possible efficient computational strategy for ensemble averages, the approach allows for assessing the accuracy of approximations commonly used in practice.

Contents

1 Introduction

Computing canonical averages is a standard task of computational materials science. Consider an atomistic system consisting of N particles, at positions $u = (u^1, \dots, u^N) \in \mathbb{R}^{3N}$. Provide this system with an energy

$$E(u) = E\left(u^1, \dots, u^N\right). \tag{1.1}$$

The finite temperature thermodynamical properties of the material are obtained from canonical ensemble averages,

$$\langle A \rangle = \frac{\displaystyle\int_{\Omega^N} A(u) \exp(-\beta E(u))\, du}{\displaystyle\int_{\Omega^N} \exp(-\beta E(u))\, du}, \tag{1.2}$$

where $\Omega \subset \mathbb{R}^3$ is the macroscopic domain where the positions u^i vary, A is the observable of interest, and $\beta = 1/(k_B T)$ is the inverse temperature [DFP02]. The denominator of (1.2) is denoted by Z and called the *partition function*. The major computational difficulty in (1.2) is of course the N-fold integrals, where N, the number of particles, is extremely large. Indeed, for integrals of the type (1.2) to be quantitatively meaningful in practice, N does not need to approach the Avogadro number, but still needs to be extremely large (10^5, say).

The three dominant computational approaches for the evaluation of (1.2) are Monte Carlo methods, Markov chains methods, and molecular dynamics methods respectively (see e.g. [CLS07] for a review of sampling methods of the canonical ensemble, along with a theoretical and numerical comparison of their performances for molecular dynamics). In the present article, we use the latter type of methods, and more precisely the overdamped Langevin dynamics (also called biased random walk). The ensemble average (1.2) is calculated as the long-time average

$$\langle A \rangle = \lim_{T \to +\infty} \frac{1}{T} \int_0^T A(u_t)\, dt \tag{1.3}$$

along the trajectory generated by the stochastic differential equation

$$du_t = -\nabla_u E(u_t)\, dt + \sqrt{2/\beta}\, dB_t. \tag{1.4}$$

It is often the case that the observable A actually does not depend on the positions u^i of *all* the atoms, but only on *some* of them. Think for instance of nanoindentation: we are especially interested in the positions of the atoms below the indenter, in the forces applied on these atoms, ... Our aim is to design a numerical method to compute the canonical averages of such observables in a more efficient way than the general strategy (1.3)-(1.4). This will also enable us to assess the validity of other approaches, as compared to ours.

The *Quasi-Continuum Method* (QCM) is a commonly used example of approaches that allow for the calculation of the specific ensemble averages discussed above. In its original version, the method was focused on the zero temperature setting. It was originally introduced in [TPO96, TOP96], and then further developed in [KO01, MTPO98, MT02, SMT+98, SMT+99, TSBK99]. It has been studied mathematically in *e.g.* [ANZEA90, AG05, AL05, BLBL05, BLBL07a, BLBL02, DL08, DL09a, DL09b, DLO09b, DLO09a, EM04, EM07, Lin03, Lin07, OS08]. See [BLBL07b, Leg09] for recent reviews. An extension of the original idea has recently been developed in [DTMP05] and carries through to the finite-temperature case, considered in the present article. See also [CC02, LSNS89] for prior studies developing ideas in the same vein.

Let us briefly detail the bottom line of coarse-graining strategies for the computation of canonical averages. For simplicity of exposition, we let the atoms vary in $\Omega = \mathbb{R}^3$. The idea is to subdivide the particles of the system into two subsets. The first subset consists of the so-called *representative atoms* (abbreviated in the QCM terminology as *repatoms*, with positions henceforth denoted by u_r). The second subset is that of atoms that are eliminated in the coarse-grained procedure. Their positions are denoted by u_c. We *assume* that the observable considered only depends on the positions u_r of the repatoms, not on those of the other atoms, u_c. More precisely, one writes

$$u = (u^1, \dots, u^N) = (u_r, u_c), \quad u_r \in \mathbb{R}^{3N_r}, \quad u_c \in \mathbb{R}^{3N_c}, \quad N = N_r + N_c,$$

and our aim is to compute (1.2) for an observable A that only depends on u_r:

$$\langle A \rangle = Z^{-1} \int_{\mathbb{R}^{3N}} A(u_r) \exp(-\beta E(u))\, du. \tag{1.5}$$

We observe that, owing to our assumption on A,

$$\int_{\mathbb{R}^{3N}} A(u_r) \exp(-\beta E(u))\, du = \int_{\mathbb{R}^{3N_r}} du_r\, A(u_r) \int_{\mathbb{R}^{3N_c}} \exp(-\beta E(u_r, u_c))\, du_c,$$

and likewise

$$Z = \int_{\mathbb{R}^{3N}} \exp(-\beta E(u))\, du = \int_{\mathbb{R}^{3N_r}} du_r \int_{\mathbb{R}^{3N_c}} \exp(-\beta E(u_r, u_c))\, du_c.$$

Introducing the coarse-grained potential (also called free energy)

$$E_{\rm CG}(u_r) := -\frac{1}{\beta} \ln \left[\int_{\mathbb{R}^{3N_c}} \exp(-\beta E(u_r, u_c))\, du_c \right], \tag{1.6}$$

the expression (1.5) reads

$$\langle A \rangle = Z_r^{-1} \int_{\mathbb{R}^{3N_r}} A(u_r) \exp(-\beta E_{\rm CG}(u_r))\, du_r, \tag{1.7}$$

with $Z_r = \int_{\mathbb{R}^{3N_r}} \exp(-\beta E_{\rm CG}(u_r))\, du_r$. Under appropriate conditions ensuring ergodicity of the system, the integral (1.7) is in turn computed from

$$\langle A \rangle = \lim_{T \to +\infty} \frac{1}{T} \int_0^T A(u_r(t))\, dt$$

with

$$du_r = -\nabla_{u_r} E_{\mathrm{CG}}(u_r)\, dt + \sqrt{2/\beta}\, dB_t. \tag{1.8}$$

Simulating the dynamics (1.8) is a less computationally demanding task than simulating (1.4), owing to the reduced dimension N_r. This simplification comes at a price: calculating the coarse-grained free energy (1.6).

Remark 1:
The present work concentrates on the computation of ensemble averages and free energies, using coarse-grained models. Practice shows that the same coarse-graining paradigm is used to simulate actual coarse-grained dynamics *at finite temperature. We will not go in this direction, as the physical relevance of the latter approach is unclear to us.*

In order to approximate the free energy (1.6), state-of-the-art finite temperature methods perform a Taylor expansion of the position of the eliminated atoms u_c. In this Taylor expansion, a linear interpolation and a harmonic approximation of the positions of the atoms are successively performed. More precisely, given the positions u_r of the repatoms, some "reference" positions $\overline{u_c}(u_r)$ of the eliminated atoms are first determined by linear interpolation between two (or more) adjacent repatoms. Then it is postulated that

$$u_c = \overline{u_c}(u_r) + \xi_c$$

where the perturbation ξ_c is small. The energy is then calculated from a Taylor expansion truncated at second order:

$$E(u_r, u_c) = E(u_r, \overline{u_c}(u_r) + \xi_c) \approx \widetilde{E}(u_r, u_c)$$

with

$$\widetilde{E}(u_r, u_c) := E(u_r, \overline{u_c}(u_r)) + \frac{\partial E}{\partial u_c}(u_r, \overline{u_c}(u_r)) \cdot \xi_c + \frac{1}{2}\xi_c \cdot \frac{\partial^2 E}{\partial u_c^2} \cdot \xi_c. \tag{1.9}$$

It follows (we skip the details of the argument and refer to the bibliography pointed out above for further details) that $E_{\mathrm{CG}}(u_r)$ is approximated by

$$E_{\mathrm{HA}}(u_r) = -\frac{1}{\beta} \ln \int_{\mathbb{R}^{3N_c}} \exp(-\beta \widetilde{E}(u_r, u_c))\, du_c, \tag{1.10}$$

which is *analytically* computable. Without such simplifying assumptions, the actual computation of E_{CG} for practical values of N_r and N_c seems impossible. The approach has proven efficient. Reportedly, it satisfactorily treats three-dimensional problems of large size. However, from the mathematical standpoint, it is an open question to evaluate the impact of the above couple of approximations (reference positions $\overline{u_c}(u_r)$ defined by a linear interpolation, followed by an harmonic expansion around these positions). The purpose of the present article is to present an approach that, in simple cases and under specific assumptions, allows for a quantitative assessment of the validity and limits of the above couple of approximations.

Our approach is based on a thermodynamic limit. It was first outlined in [Pat09] for the special case of harmonic interactions. The approach is *exact* in the limit of an infinite number of eliminated atoms, and therefore valid when this number N_c is large as compared to the number N_r of representative atoms that are kept explicit in the coarse-grained model. This regime, after all, is the regime that all effective coarse-graining strategies should target. In short, the consideration of the asymptotic limit $N_c \to +\infty$ makes tractable a computation which is not tractable for finite N_c (unless simplifications, as those mentioned above, are performed). We do not claim for originality in our theoretical considerations on the thermodynamic limit of the free energy of atomistic systems. We provide them here for consistency. However, our specific use of such theoretical considerations as a computational strategy for approximating coarse-grained ensemble averages in computational materials science seems, to the best of our knowledge, new. We were not able to find any comparable endeavor in the existing literature we have access to.

Let us conclude this introduction by briefly describing our approach. Assume for simplicity that there is only one repatom: $N_r = 1$. Our idea to compute $\langle A \rangle$ in (1.5) is to change variables, that is introduce $y = (y_1, \ldots, y_N) = \Phi(u)$, and recast (1.5) as

$$\langle A \rangle = \int_{\mathbb{R}^{3N}} A\left(\frac{1}{N}\sum_{i=1}^{N} y_i\right) \nu(y)\, dy$$

for some probability density $\nu(y)$ (see equation (2.5) below for an explicit example). We next recognize $\langle A \rangle$ as the expectation value $\mathbb{E}\left[A\left(\frac{1}{N}\sum_{i=1}^{N} Y_i\right)\right]$, where $Y = (Y_1, \ldots, Y_N)$ are random variables distributed according to the probability ν. A Law of Large Numbers provides the limit of $\langle A \rangle$ when $N \to +\infty$ (which corresponds to $N_c \to +\infty$, since $N_r = 1$). The rate of convergence may also be evaluated using the Central Limit Theorem.

Note that the above approach bypasses the calculation of the free energy $E_{\rm CG}$ to compute the ensemble average (1.5). Our strategy is hence different from the one we described above, which is based on using the formula (1.7).

It is also interesting to try and evaluate $E_{\rm CG}$ in the same regime. First, it is to be remarked that $E_{\rm CG}$ scales linearly with the number N_c of eliminated atoms. The relevant quantity is hence the free energy *per particle*

$$F_\infty(u_r) := \lim_{N_c \to +\infty} \frac{1}{N_c} E_{\rm CG}(u_r). \tag{1.11}$$

We now observe that $N_c F_\infty$ is not necessarily a good approximation of $E_{\rm CG}$, for large N_c, even if F_∞ is a good approximation of $E_{\rm CG}/N_c$. It is not clear to us how to use F_∞, or the probability measure $Z_{N_c}^{-1} \exp(-\beta N_c F_\infty)$, to compute in an efficient manner an approximation of the average $\langle A \rangle$ (see Remark 8 below). Nevertheless, computing F_∞ turns out to be also interesting, but for a different reason. This energy is indeed related to the coarse-grained constitutive law of the material at finite temperature, as we explain at the end of Section 2.2 (see [Der07, Oll07] for related approaches).

We develop our approach in the one-dimensional setting, for simple cases of pair interactions. We first consider nearest neighbor (NN) interactions. In this case, we develop a computational strategy to approximate ensemble averages (see Section 2.1), and we next address the computation of free energies (see Section 2.2). Numerical considerations are collected in Section 2.3.

We next turn to next-to-nearest neighbor interactions, traditionally abbreviated as NNN. For this model, we focus on the computation of ensemble averages (see Section 3.1). As explained in Section 3.2, more complicated types of interaction potentials and "essentially one-dimensional systems" (including polymer chains) may be treated likewise, although we do not pursue in this direction.

A similar interpretation of ensemble averages as the one presented here, using a Markov chain formalism, *should* lead to an analogous strategy for two-dimensional systems. Some preliminary developments, not included in the present article, already confirm this. However, definite conclusions are yet to be obtained, both on the formal validity of the approach and on the best possible numerical efficiency accomplished. The fact that the two-dimensional case is much more difficult than the one-dimensional case is corroborated by the literature on this subject: only very simple cases, such as spin systems, or harmonic interactions (with zero equilibrium length) are known to have explicit solutions in this context (see the reviews [Bax82, Pre09]). We therefore prefer to postpone considerations on the two-dimensional situation until a future publication.

2 The nearest neighbour (NN) case

As mentioned in the introduction, our approach is based on the asymptotic limit $N \longrightarrow +\infty$. We therefore first rescale the problem with the inter-atomic distance h, such that $Nh = L = 1$ (see Remark 2 below). The atomistic energy (1.1) in the *rescaled* NN case writes

$$E\left(u^1, \dots, u^N\right) = \sum_{i=1}^{N} W\left(\frac{u^i - u^{i-1}}{h}\right), \tag{2.1}$$

where $W : \mathbb{R} \to \mathbb{R}$ is an inter-atomic potential (see (2.37) for a precise example, and theorems below for precise assumptions on W). We now impose $u^0 = 0$ to avoid translation invariance, and consider that only atoms 0 and N are repatoms, while all the other atoms $i = 1, \dots, N-1$ are eliminated in the coarse-graining procedure (see Figure 2.1). Our argument can be straightforwardly adapted to treat the case of $N_r > 2$ repatoms (see Figure 2.2 and the end of Section 2.1).

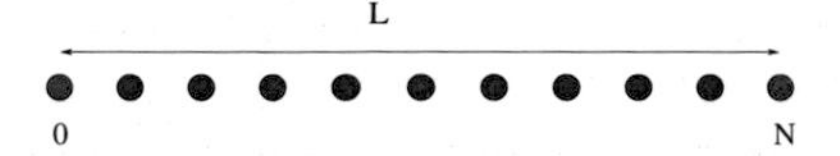

Figure 2.1: We isolate a segment between two consecutive repatoms (in red). All atoms in-between (in blue) are eliminated in the coarse-graining procedure.

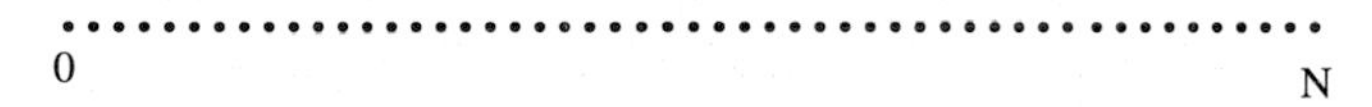

Figure 2.2: The repatoms (in red) are explicitly treated, all other atoms (in blue) being eliminated in the coarse-graining procedure.

In this simple situation, the average (1.5) reads

$$\langle A \rangle_N = Z^{-1} \int_{\mathbb{R}^N} A\left(u^N\right) \exp\left(-\beta \sum_{i=1}^{N} W\left(\frac{u^i - u^{i-1}}{h}\right)\right) du^1 \dots du^N, \tag{2.2}$$

where we have explicitly mentioned the dependence of $\langle A \rangle$ upon N using a subscript.

Remark 2:
Instead of the energy (2.1)*, we can work with the energy*

$$\widetilde{E}\left(u^1, \dots, u^N\right) = \sum_{i=1}^{N} W\left(u^i - u^{i-1}\right),$$

for an inter-atomic potential W that reaches its minimum at a value $\bar{x}$ independent of h (say $\bar{x} = 1$). In that case, as $N \to \infty$, the mean size of the system diverges, and we thus expect averages of the type (1.5) *to diverge as well: for instance,*

$$\lim_{N \to \infty} \widetilde{Z}^{-1} \int u^N \exp(-\beta \widetilde{E}(u))\, du = \infty.$$

With that scaling, the relevant quantity to consider is

$$\widetilde{Z}^{-1} \int A\left(\frac{u^N}{N}\right) \exp(-\beta \widetilde{E}(u))\, du$$

which is exactly (2.2).

We introduce

$$y_i := \frac{u^i - u^{i-1}}{h}, \quad i = 1, \dots, N, \tag{2.3}$$

and next remark that

$$u^N = h \sum_{i=1}^{N} y_i = \frac{1}{N} \sum_{i=1}^{N} y_i. \tag{2.4}$$

Thus, the average (2.2) reads

$$\langle A \rangle_N = Z^{-1} \int_{\mathbb{R}^N} A\left(\frac{1}{N}\sum_{i=1}^{N} y_i\right) \exp\left(-\beta \sum_{i=1}^{N} W(y_i)\right) dy^1 \dots dy^N, \qquad (2.5)$$

where now $Z = \int_{\mathbb{R}^N} \exp\left(-\beta \sum_{i=1}^{N} W(y_i)\right) dy^1 \dots dy^N$.

Remark 3:
In (2.2), we let the variables u^i vary on the whole real line. We do not constrain them to obey $u^{i-1} \leq u^i$, which encodes the fact that nearest neighbors remain nearest neighbors. The argument provided here and below carries through when this constraint is accounted for: we just need to replace the interaction potential W by

$$W_c(y) = \begin{cases} W(y) & \text{when} \quad y \geq 0 \\ +\infty & \text{otherwise.} \end{cases} \qquad (2.6)$$

Likewise, we could also impose that all the u^i stay in a given macroscopic segment. If they are ordered increasingly, it is enough to impose this constraint on u^0 and u^N. This is again a simple modification of our argument.

2.1 Limit of the average

It is evident from the expression (2.5) that

$$\langle A \rangle_N = \mathbb{E}\left[A\left(\frac{1}{N}\sum_{i=1}^{N} Y_i\right)\right]$$

for independent identically distributed (i.i.d.) random variables Y_i, sharing the law $z^{-1}\exp(-\beta W(y))dy$, with $z = \int_{\mathbb{R}} \exp(-\beta W(y))dy$. A simple computation thus gives the following result:

Theorem 1:
Assume that $A : \mathbb{R} \longrightarrow \mathbb{R}$ is continuous, that for some $p \geq 1$, there exists a constant $C > 0$ such that

$$\forall y \in \mathbb{R}, \quad |A(y)| \leq C(1 + |y|^p), \qquad (2.7)$$

and that

$$\int_{\mathbb{R}} (1 + |y|^p) \exp(-\beta W(y))\, dy < +\infty. \qquad (2.8)$$

Introduce y^* and σ defined by

$$y^* = z^{-1} \int_{\mathbb{R}} y \ \exp(-\beta W(y))\, dy$$
$$\sigma^2 = z^{-1} \int_{\mathbb{R}} (y - y^*)^2 \ \exp(-\beta W(y))\, dy, \qquad (2.9)$$

with $z = \int_{\mathbb{R}} \exp(-\beta W(y))\, dy$*. Then,*

$$\lim_{N \to +\infty} \langle A \rangle_N = A(y^*). \tag{2.10}$$

In addition, if A *is* C^2 *and if* (2.7)-(2.8) *hold with* $p = 2$, *then* $\sigma < \infty$ *and*

$$\langle A \rangle_N = \langle A \rangle_N^{\text{approx}} + o\left(\frac{1}{N}\right), \tag{2.11}$$

with

$$\langle A \rangle_N^{\text{approx}} := A(y^*) + \frac{\sigma^2}{2N} A''(y^*). \tag{2.12}$$

The proof of (2.10) is a direct application of the Law of Large Numbers, and that of (2.11) is an application of the Central Limit Theorem. We skip them. The following considerations, for more regular observables A, indeed contain the ingredients for proving (2.10)-(2.11), simply by truncating the expansion at first order.

Remark 4:
Note that other growth assumptions on A *are possible, along with corresponding assumptions on* W*. For instance, if* A *satisfies* $|A(y)| \leq C \exp(|y|)$ *for some* C *and if* W *is such that* $\exp(|y| - \beta W(y))$ *is integrable, then* (2.10) *holds.*

If A is more regular than stated in Theorem 1, then it is of course possible to proceed further in the expansion of $\langle A \rangle_N$ in powers of $1/N$. Indeed, assume for instance that A is C^6, that $A^{(6)}$ is globally bounded and that (2.7)-(2.8) hold with $p = 6$. Then

$$\begin{aligned} A\left(\frac{1}{N}\sum_{i=1}^N Y_i\right) &= A\left(y^* + \frac{1}{N}\sum_{i=1}^N D_i\right), \\ &= A(y^*) + A'(y^*)\frac{1}{N}\sum_{i=1}^N D_i + \frac{1}{2}A''(y^*)\left(\frac{1}{N}\sum_{i=1}^N D_i\right)^2 \\ &\quad + \frac{1}{6}A^{(3)}(y^*)\left(\frac{1}{N}\sum_{i=1}^N D_i\right)^3 + \frac{1}{24}A^{(4)}(y^*)\left(\frac{1}{N}\sum_{i=1}^N D_i\right)^4 \\ &\quad + \frac{1}{5!}A^{(5)}(y^*)\left(\frac{1}{N}\sum_{i=1}^N D_i\right)^5 + \frac{1}{6!}A^{(6)}(\xi)\left(\frac{1}{N}\sum_{i=1}^N D_i\right)^6, \end{aligned}$$

where $D_i = Y_i - y^*$ and ξ lies between y^* and $(1/N)\sum Y_i$. We now take the expectation value of this equality. Let us introduce

$$\begin{aligned} \mathcal{A}_N &:= A(y^*) + \frac{1}{2}A''(y^*)\frac{1}{N}\mathbb{E}\left(D_1^2\right) + \frac{1}{6}A^{(3)}(y^*)\frac{1}{N^2}\mathbb{E}\left(D_1^3\right) \\ &\quad + \frac{1}{24}A^{(4)}(y^*)\left(\frac{1}{N^3}\mathbb{E}\left(D_1^4\right) + \frac{N-1}{N^3}\left(\mathbb{E}(D_1^2)\right)^2\right) \\ &\quad + \frac{1}{5!}A^{(5)}(y^*)\left(\frac{1}{N^4}\mathbb{E}\left(D_1^5\right) + \frac{N-1}{N^4}\mathbb{E}(D_1^2)\mathbb{E}(D_1^3)\right). \end{aligned} \tag{2.13}$$

Then

$$|\langle A\rangle_N - \mathcal{A}_N| \leq \frac{1}{6!}\,\|A^{(6)}\|_{L^\infty}\;\mathbb{E}\left[\left(\frac{1}{N}\sum_{i=1}^N D_i\right)^6\right]. \tag{2.14}$$

We now use the fact that any i.i.d. variables D_i with mean value zero satisfy the following bounds:

$$\forall p \in \mathbb{N}, \quad \exists C_p > 0, \quad \left|\mathbb{E}\left[\left(\frac{1}{N}\sum_{i=1}^N D_i\right)^p\right]\right| \leq \begin{cases} \dfrac{C_p}{N^{\frac{p}{2}}} & \text{if } p \text{ is even;} \\ \dfrac{C_p}{N^{\frac{p+1}{2}}} & \text{if } p \text{ is odd.} \end{cases} \tag{2.15}$$

This is proved by developing the power p of the sum, and then using the fact that the variables are i.i.d and have mean value zero. We hence infer from (2.13), (2.14) and (2.15) that

$$\langle A\rangle_N = A(y^*) + \frac{\sigma^2}{2N}A''(y^*) + \frac{1}{N^2}\left(\frac{m_3}{6}A^{(3)}(y^*) + \frac{\sigma^4}{24}A^{(4)}(y^*)\right) + O\left(\frac{1}{N^3}\right),$$

where σ is defined by (2.9) and

$$m_3 = z^{-1}\int_{\mathbb{R}} (y-y^*)^3 \exp\left(-\beta W(y)\right) dy. \tag{2.16}$$

More generally, it is possible to expand $\langle A\rangle_N$ at any order in $1/N$, provided that A is sufficiently smooth and $\exp(-\beta W)$ sufficiently small at infinity. In view of the bounds (2.15), we can see that using a Taylor expansion of order $2p$ around y^* for A gives an expansion of $\langle A\rangle_N$ of order p.

The practical consequence of Theorem 1 is that, for computational purposes, we may take the approximation

$$\langle A\rangle_N \approx A\left(z^{-1}\int_{\mathbb{R}} y\,\exp(-\beta W(y))\,dy\right). \tag{2.17}$$

As pointed out above, it is possible to improve this approximation if necessary by expanding further in powers of $1/N$.

We conclude this section by showing that our consideration of a single "segment" carries through to the case when there are 3 repatoms, of respective index 0, M_1 and $M_1 + M_2$, with $M_1 h = L_1$, $M_2 h = L_2$, $Nh = L = 1$ (see Figure 2.3). The average to compute is

$$\langle A \rangle_N = Z^{-1} \int_{\mathbb{R}^N} A\left(u^{M_1}, u^{M_1+M_2}\right) \exp\left(-\beta \sum_{i=1}^{N} W\left(\frac{u^i - u^{i-1}}{h}\right)\right) du^1 \ldots du^N.$$

In the regime $h \to 0$, $N, M_1, M_2 \to +\infty$ with M_1/N and M_2/N fixed, we have, using similar arguments,

$$\lim_{N \to +\infty} \langle A \rangle_N = A(L_1 y^*, (L_1 + L_2) y^*).$$

The generalization to $N_r > 3$ repatoms, in the appropriate asymptotic regime, easily follows.

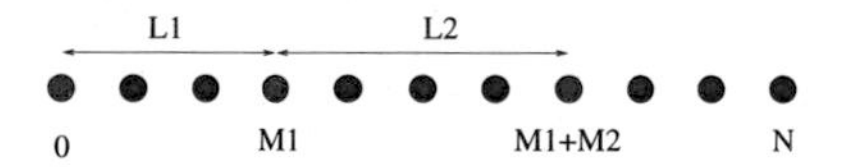

Figure 2.3: When considering two consecutive segments -or more-, the argument may be readily adapted. See the text.

Remark 5 (The small temperature limit):
It is interesting here to consider the small temperature limit of the above expansion, that is, the limit $\beta \to +\infty$. In such a case, using the Laplace method (see [BO78]), it is possible to compute the limit of the terms that appear in the expansion (2.12) of $\langle A \rangle_N$. We give as an example the first and the second terms:

$$A(y^*) = A(a) + O\left(\frac{1}{\beta}\right), \qquad \frac{\sigma^2}{2} A''(y^*) = \frac{1}{2\beta} \frac{A''(a)}{W''(a)} + O\left(\frac{1}{\beta^2}\right),$$

where a is the point where W attains its minimum (in this remark, we assume for simplicity that W attains its minimum at a unique point). Hence,

$$\langle A \rangle_N = \left[A(a) + O\left(\frac{1}{\beta}\right)\right] + \frac{1}{N}\left[\frac{1}{2\beta}\frac{A''(a)}{W''(a)} + O\left(\frac{1}{\beta^2}\right)\right] + O\left(\frac{1}{N^2}\right).$$

Now, it is possible to recover these terms by expanding the energy E around the equilibrium configuration corresponding to $y_i = a$. Indeed, if we assume that $W(y) = W''(a)\,(y-a)^2/2$ in (2.2), then a simple explicit computation gives

$$\langle A \rangle_N = A(a) + \frac{1}{2N\beta W''(a)} A''(a) + O\left(\frac{1}{\beta^2 N^2}\right).$$

Hence, expanding the first terms of (2.12) *in powers of $1/\beta$ for large β gives an expansion that agrees with that obtained using a harmonic approximation of the energy. This provides a quantitative evaluation of the latter approach in this asymptotic regime.*

2.2 Limit of the free energy

We now look for a more demanding result. For clarity, let us return to the general coarse-grained average (1.7), which of course equals (2.2) and (2.5) in our simple NN case. Instead of searching for the limit of the average $\langle A \rangle$ for large N_c, we now look for the limit of the free energy per particle (see (1.6) and (1.11)). We discuss below (see Remark 8 and the end of this section) the interpretation of that quantity.

In the present section, u_r is in fact equal to u^N (the right end atom) since atom 0, although a repatom, is fixed to avoid translation invariance: $u^0 = 0$. Thus we wish to identify the behavior for N large of

$$E_{\mathrm{CG}}\left(u^N\right) = -\frac{1}{\beta} \ln \left[\int_{\mathbb{R}^{N-1}} \exp\left(-\beta E\left(u^1, \ldots, u^N\right)\right) du^1 \ldots u^{N-1} \right]. \quad (2.18)$$

Note that E_{CG} is the free energy corresponding to integrating out $N-1$ variables. From Thermodynamics, it is expected that E_{CG} scales linearly with N. This is confirmed by the consideration of the harmonic potential $W(x) = \frac{k}{2}(x-a)^2$, for which $E_{\mathrm{CG}}\left(u^N\right) = \frac{kN}{2}\left(u^N - a\right)^2 + C(N, \beta, k)$, where

$$C(N, \beta, k) = \frac{1}{\beta}\left(N - \frac{1}{2}\right) \ln N - \frac{N-1}{2\beta} \ln\left(\frac{2\pi}{\beta k}\right)$$

does not depend on u_N (see the details in [Pat09]). Therefore, we introduce the free energy per particle

$$F_N(x) := \frac{1}{N} E_{\mathrm{CG}}(x),$$

so that

$$\langle A \rangle_N = Z_r^{-1} \int_{\mathbb{R}} A\left(u^N\right) \exp\left(-\beta N F_N\left(u^N\right)\right) du^N. \quad (2.19)$$

The limit behavior of F_N is provided by the Large Deviations Principle. This claim is made precise in the following theorem.

Theorem 2:
Assume that the potential W satisfies

$$\forall \xi \in \mathbb{R}, \quad \int_{\mathbb{R}} \exp\left(\xi y - \beta W(y)\right) dy < +\infty, \quad (2.20)$$

and $\exp(-\beta W) \in H^1(\mathbb{R} \setminus \{0\})$. Then the limit behavior of F_N is given by the following Legendre transform:

$$\lim_{N \to +\infty} \left(F_N(x) + \frac{1}{\beta} \ln \frac{z}{N} \right) = F_\infty(x) \quad (2.21)$$

with

$$F_\infty(x) := \frac{1}{\beta} \sup_{\xi \in \mathbb{R}} \left(\xi x - \ln \left[z^{-1} \int_{\mathbb{R}} \exp(\xi y - \beta W(y))\, dy \right] \right) \quad (2.22)$$

and $z = \int_{\mathbb{R}} \exp(-\beta W(y))\, dy$.

Remark 6:
The assumption $\exp(-\beta W) \in H^1(\mathbb{R}\backslash\{0\})$ *allows for* W *to be piecewise continuous, with discontinuity at the origin. This in particular allows us to deal with the type of potentials mentioned in Remark 3.*

Proof. Let us first rewrite the free energy $F_N(x)$ as follows:

$$
\begin{aligned}
F_N(x) &= -\frac{1}{\beta N} \ln \left[\int_{\mathbb{R}^{N-1}} \exp\left(-\beta \sum_{i=1}^{N-1} W\left(\frac{u^i - u^{i-1}}{h} \right) \right.\right. \\
&\qquad \left.\left. -\beta W\left(\frac{x - u^{N-1}}{h} \right) \right) du^1 \dots du^{N-1} \right] \\
&= -\frac{N-1}{\beta N} \ln h - \frac{1}{\beta N} \ln \left[\int_{\mathbb{R}^{N-1}} \exp\left(-\beta W \left(Nx - \sum_{i=1}^{N-1} y_i \right) \right.\right. \\
&\qquad \left.\left. -\beta \sum_{i=1}^{N-1} W(y_i) \right) dy_1 \dots dy_{N-1} \right] \\
&= -\frac{1}{\beta} \ln h - \frac{1}{\beta} \ln z - \frac{1}{\beta N} \ln \mu_N(x),
\end{aligned}
$$

where μ_N is the law of the random variable $(1/N)\sum_{i=1}^N Y_i$ and Y_i is a sequence of i.i.d. random variables with law $\mu(y) = z^{-1}\exp(-\beta W(y))$. Observe that

$$
\mu_N(x) = \frac{N}{z^N} \int_{\mathbb{R}^{N-1}} \exp\left(-\beta W \left(Nx - \sum_{i=1}^{N-1} y_i \right) - \beta \sum_{i=1}^{N-1} W(y_i) \right) dy_1 \dots dy_{N-1}.
$$

We also have

$$
\mu_N(x) = N\ \mu^{*N}(Nx),
$$

where μ^{*N} denotes the $(N-1)$-fold convolution product of μ ($\mu^{*2} = \mu * \mu$).

The sequence of measures μ_N satisfies a large deviations property (see for instance [Ell85a, Ell85b, Ell95, Var84]). We are going to use it in order to compute the limit of $\frac{1}{N}\ln \mu_N$. We first prove a lower bound, which is a simple consequence of the results of [Var84]. The upper bound is more involved: we need to reproduce the corresponding proof of [Var84], and use a refined version of the Central Limit Theorem (see [LT95]).

We introduce the function

$$
G_N(x) = -\frac{1}{\beta N} \ln \mu_N(x), \tag{2.23}
$$

which satisfies, in view of the above computation,

$$
F_N(x) = -\frac{1}{\beta} \ln \frac{z}{N} + G_N(x). \tag{2.24}
$$

First step: lower bound. We write

$$\mu_{N+1}(x) = (N+1) \int_{\mathbb{R}} \mu(N(x-t)+x)\, \mu_N(t)\, dt. \tag{2.25}$$

Let us define

$$J_N^x(t) = -\frac{1}{N} \ln \mu\left(N(x-t)+x\right).$$

This function clearly satisfies the following convergence:

$$\liminf_{u \to t, N \to +\infty} J_N^x(u) = J_\infty^x(t) := \begin{cases} +\infty & \text{if } t \neq x, \\ 0 & \text{if } t = x. \end{cases}$$

Hence, we may apply Theorem 2.3 of [Var84], which implies that

$$\begin{aligned} &\liminf_{N \to +\infty} \left(-\frac{1}{N} \ln \int_{\mathbb{R}} \exp\left(-N J_N^x(t)\right) \mu_N(t) dt \right) \\ &\geq \inf_{t \in \mathbb{R}} \left(J_\infty^x(t) + \beta F_\infty(t)\right) = \beta F_\infty(x), \end{aligned} \tag{2.26}$$

where F_∞ is defined by (2.22). Since the left-hand side of (2.26) is equal to $\frac{\beta(N+1)}{N} G_{N+1}(x) + \frac{\ln(N+1)}{N}$, we infer from the above bound that

$$\liminf_{N \to +\infty} G_N(x) \geq F_\infty(x). \tag{2.27}$$

Second step: upper bound. We now aim at bounding G_N from above. We recall that the function we maximize in (2.22) is concave, so there exists a unique $\xi_x \in \mathbb{R}$ such that

$$F_\infty(x) = \frac{1}{\beta} \left(\xi_x x - \ln \left[z^{-1} \int_{\mathbb{R}} \exp\left(\xi_x y - \beta W(y)\right) dy \right] \right).$$

The Euler-Lagrange equation of the maximization problem implies

$$x = \frac{\displaystyle\int_{\mathbb{R}} y \exp\left(\xi_x y - \beta W(y)\right) dy}{\displaystyle\int_{\mathbb{R}} \exp\left(\xi_x y - \beta W(y)\right) dy}. \tag{2.28}$$

We introduce the notations

$$\widetilde{\mu}(t) = \frac{\exp(\xi_x t - \beta W(t))}{\displaystyle\int_{\mathbb{R}} \exp(\xi_x t - \beta W(t))\, dt} \quad \text{and} \quad M(\xi) = z^{-1} \int_{\mathbb{R}} \exp(\xi t - \beta W(t))\, dt,$$

and compute

$$\begin{aligned}\mu_N(x) &= N\int_{\mathbb{R}^{N-1}}\mu\left(Nx-\sum_{i=1}^{N-1}y_i\right)\mu(y_1)\dots\mu(y_{N-1})\,dy_1\dots dy_{N-1}\\ &= N\ M(\xi_x)^{N-1}\int_{\mathbb{R}^{N-1}}\mu\left(Nx-\sum_{i=1}^{N-1}y_i\right)\exp\left(-\xi_x\sum_{i=1}^{N-1}y_i\right)\\ &\qquad\times\widetilde{\mu}(y_1)\dots\widetilde{\mu}(y_{N-1})\,dy_1\dots dy_{N-1}\\ &\geq N\ M(\xi_x)^{N-1}\int_{|Nx-\sum y_i|\leq\delta}\mu\left(Nx-\sum_{i=1}^{N-1}y_i\right)\\ &\qquad\times\exp\left(-\xi_x\sum_{i=1}^{N-1}y_i\right)\widetilde{\mu}(y_1)\dots\widetilde{\mu}(y_{N-1})\,dy_1\dots dy_{N-1}\\ &\geq N\ M(\xi_x)^{N-1}\left(\inf_{[-\delta,\delta]}\mu\right)\exp(-\xi_xNx-|\xi_x|\delta)\\ &\qquad\times\int_{|Nx-\sum y_i|\leq\delta}\widetilde{\mu}(y_1)\dots\widetilde{\mu}(y_{N-1})\,dy_1\dots dy_{N-1},\end{aligned}$$

for any $\delta>0$. Hence,

$$\begin{aligned}G_N(x)\leq -\frac{1}{\beta N}\ln N-\frac{N-1}{\beta N}\ln(M(\xi_x))+\frac{\xi_x x}{\beta}+|\xi_x|\frac{\delta}{\beta N}\\ -\frac{1}{\beta N}\ln\left(\inf_{[-\delta,\delta]}\mu\right)-\frac{1}{\beta N}\ln\mathbb{P}\left(\left|\frac{1}{N}\sum_{i=1}^{N-1}Y_i-x\right|\leq\frac{\delta}{N}\right),\end{aligned}\qquad(2.29)$$

where the random variables Y_i are i.i.d. of law $\widetilde{\mu}$. The equation (2.28) implies that $\mathbb{E}(Y_i)=x$. According to the hypotheses on W, we have $\widetilde{\mu}\in H^1(\mathbb{R}\setminus\{0\})$, hence we may apply Theorem 5.1 of [LT95]. It implies that the law θ_N of the variable $\left(\sum_{i=1}^N Y_i-Nx\right)/\sqrt{N}$ converges in $H^1(\mathbb{R})$ to some normal law. In particular, we have convergence in L^∞, hence

$$\mathbb{P}\left(\left|\frac{1}{N}\sum_{i=1}^{N-1}Y_i-x\right|\leq\frac{\delta}{N}\right)=\int_{\frac{x-\delta}{\sqrt{N-1}}}^{\frac{x+\delta}{\sqrt{N-1}}}\theta_{N-1}(t)dt\geq\frac{2\gamma\delta}{\sqrt{N-1}},$$

for N large enough, where $\gamma>0$ does not depend on N. Inserting this inequality into (2.29), we find

$$\begin{aligned}G_N(x)\leq -\frac{1}{\beta N}\ln N-\frac{N-1}{\beta N}\ln(M(\xi_x))+\frac{\xi_x x}{\beta}+|\xi_x|\frac{\delta}{\beta N}\\ -\frac{1}{\beta N}\ln\left(\inf_{[-\delta,\delta]}\mu\right)-\frac{1}{\beta N}\ln\left(\frac{2\gamma\delta}{\sqrt{N-1}}\right).\end{aligned}\qquad(2.30)$$

Hence,

$$\limsup_{N\to+\infty} G_N(x) \le -\frac{1}{\beta}\ln(M(\xi_x)) + \frac{\xi_x x}{\beta},$$

which implies, according to the definition of M and ξ_x, that

$$\limsup_{N\to+\infty} G_N(x) \le F_\infty(x). \tag{2.31}$$

Estimates (2.27) and (2.31) imply $\lim_{N\to+\infty} G_N(x) = F_\infty(x)$. In view of (2.24), this implies (2.21). □

Remark 7 (The small temperature limit):
As in Remark 5, it is possible to compute the expansion of $F_\infty(x)$ as $\beta \to +\infty$. Using the Laplace method, and assuming that W is convex, one finds that

$$F_\infty(x) = W(x) + \frac{1}{2\beta}\ln W''(x) + O\left(\frac{1}{\beta^2}\right).$$

Let us now consider another strategy to find an approximation of F_N. In the spirit of the Quasi-Continuum Method, we expand $E(u^1,\ldots,u^N)$ around the equilibrium configuration $\overline{u}^i = iu^N/N$, for a given u^N. More precisely, we set $u^i = \overline{u}^i + \xi_i$, assume that ξ_i is small, and expand the energy at second order with respect to ξ_i, as explained in the Introduction. This yields the approximate energy $\widetilde{E}$ defined by (1.9), that we next insert in (2.18) (as we did in (1.10)). Due to the harmonic approximation, the resulting coarse-grained energy, that we denote $E_{\rm HA}$, is analytically computable and reads

$$E_{\rm HA}(x) = NW(x) + \frac{N-1}{2\beta}\ln W''(x) + \frac{N-1}{2\beta}\ln\frac{\beta}{2\pi} + \frac{1}{2\beta}\ln N. \tag{2.32}$$

Hence,

$$F_{\rm HA}(x) := \lim_{N\to+\infty}\frac{1}{N}E_{\rm HA}(x) = W(x) + \frac{1}{2\beta}\ln W''(x) + \frac{1}{2\beta}\ln\frac{\beta}{2\pi}. \tag{2.33}$$

Thus, up to an additive constant, $F_{\rm HA}(x)$ corresponds to the first-order approximation (in powers of $1/\beta$) of $F_\infty(x)$.

Slightly improving the proof of Theorem 2 above, it is possible to prove the convergence of the derivative of the free energy, a quantity which is indeed practically relevant (e.g. for the simulation of (1.8)):

Corollary 1:
Assume that the hypotheses of Theorem 2 are satisfied. Then, we have

$$F_N(x) + \frac{1}{\beta}\ln\left(\frac{z}{N}\right) \longrightarrow F_\infty(x) \quad \textit{in} \quad L^p_{\rm loc}, \quad \forall p \in [1,+\infty).$$

In particular, this implies that F_N' converges to F_∞' in $W^{-1,p}_{\rm loc}$.

Proof. According to Theorem 2, we already know the pointwise convergence of $G_N(x) = F_N(x) + \beta^{-1} \ln(z/N)$. We therefore only need to prove that G_N is bounded in L^∞_{loc} to prove our claim.

Lower bound: We go back to (2.25), and point out that $\mu \leq 1/z$. Hence,

$$\mu_{N+1}(x) \leq \frac{N+1}{z} \int_{\mathbb{R}} \mu_N = \frac{N+1}{z},$$

which implies, using (2.23), that

$$G_{N+1}(x) \geq -\frac{1}{\beta(N+1)} \ln \frac{N+1}{z},$$

which is bounded from below independently of N.

Upper bound: We return to (2.30), and notice that according to the definition of ξ_x, the function $x \mapsto \xi_x$ is continuous. In addition, the constant γ in (2.30) is a continuous function of ξ_x. Therefore, (2.30) provides an upper bound on G_N.

As a conclusion, G_N is bounded in L^∞_{loc}, which allows to conclude. □

Remark 8:
Considering the above theoretical results, it could be tempting to approach the average (2.19), *that is,*

$$\langle A \rangle_N = Z_r^{-1} \int_{\mathbb{R}} A\left(u^N\right) \exp\left(-\beta N F_N\left(u^N\right)\right) du^N,$$

by

$$Z_\infty^{-1} \int_{\mathbb{R}} A\left(u^N\right) \exp\left(-\beta N F_\infty\left(u^N\right)\right) du^N, \tag{2.34}$$

with $Z_\infty = \int_{\mathbb{R}} \exp\left(-\beta N F_\infty\left(u^N\right)\right) du^N$. *Note that F_N has been replaced by F_∞ in the exponential factor. This strategy is not efficient since this approximation does not provide the expansion (2.11)-(2.12) of $\langle A \rangle_N$ in powers of* $1/N$. *Indeed, it is possible to use the Laplace method to compute the expansion of* (2.34) *as* $N \to +\infty$. *It reads*

$$A(y^*) + \frac{1}{2N}\left(\sigma^2 A''(y^*) + \frac{m_3}{\sigma^2} A'(y^*)\right) + o\left(\frac{1}{N}\right),$$

where σ is defined by (2.9) and m_3 is defined by (2.16). *This expansion coincides with (2.11)-(2.12) only for the first term, that is $A(y^*)$. The second one differs, unless $m_3\ A'(y^*) = 0$.*

To improve the approximation (2.34), *one may use the precised large deviations principle (see [DZ93, Theorem 3.7.4] or [BRR60]). In such a case, one replaces* (2.34) *by*

$$\widetilde{Z}_\infty^{-1} \int_{\mathbb{R}} A\left(u^N\right) \sqrt{F''_\infty\left(u^N\right)}\ \exp\left(-\beta N F_\infty\left(u^N\right)\right) du^N, \tag{2.35}$$

with $\widetilde{Z}_\infty = \int_{\mathbb{R}} \sqrt{F''_\infty(u^N)} \exp\left(-\beta N F_\infty(u^N)\right) du^N$. *This quantity is well-defined since* F_∞ *is a convex function. Then it is seen that the expansion of* (2.35) *in powers of* $1/N$ *agrees with (2.11)-(2.12) up to the second order term. Note however that using* (2.35) *leads to a much more expensive computation than using* (2.12), *since it requires the evaluation of* F_∞ *and its second derivative.*

The above convergence of the free energy F_N is useful for the computation of the free energy of a chain of atoms with a *prescribed* length. Indeed, consider a chain on which we impose

$$u^N = \ell,$$

for a fixed ℓ. We aim at computing the free energy F_N as a function of ℓ, in the limit $N \to +\infty$. We have

$$F_N(\ell) = -\frac{1}{\beta N} \ln \left[\int_{\mathbb{R}^{N-1}} \exp\left(-\beta \sum_{i=1}^{N} W\left(\frac{u^i - u^{i-1}}{h}\right)\right) du^1 \dots du^{N-1} \right],$$

where $u^N = \ell$. The limit of F_N is provided by Theorem 2.

Another interest of the approach is to provide an approximation of $F'_N(\ell)$, a quantity related to the constitutive law of the material under consideration, at the *finite temperature* $1/\beta$. Indeed, note that

$$F'_N(\ell) = \langle A_N \rangle^\ell_N, \tag{2.36}$$

where $\langle \cdot \rangle^\ell_N$ is the average with respect to the Gibbs measure $d\mu_\ell(u^1, \dots, u^{N-1})$ associated to the energy $\sum_{i=1}^{N-1} W\left(\frac{u^i - u^{i-1}}{h}\right) + W\left(\frac{\ell - u^{N-1}}{h}\right)$ (recall that $h = 1/N$), and the observable A_N is defined by

$$A_N\left(u^1, \dots, u^{N-1}\right) = W'\left(N\left(\ell - u^{N-1}\right)\right).$$

Note that $W'\left(N\left(\ell - u^{N-1}\right)\right)$ can be interpreted as the force in the spring between atom $N-1$ and N, when the position of the latter is fixed at the value ℓ. Hence $F'_N(\ell)$ can be interpreted as the average force between atoms $N-1$ and N, when the position of atom N is prescribed at $u^N = \ell$, and the relation $\ell \mapsto F'_N(\ell)$ can be considered as the constitutive relation (at a given temperature) of the chain, providing the stress $F'_N(\ell)$ as a function of the strain ℓ. Corollary 1 provides the convergence of $F'_N(\ell)$ to $F'_\infty(\ell)$.

Remark 9:
Note that, in (2.36), the observable A_N *depends on* N. *Hence, the results of Section 2.1 (obtained using the Law of Large Numbers and not involving the Large Deviations Principle) do not apply to compute the large* N *limit of* $\langle A_N \rangle^\ell_N$. *In addition, the Gibbs measure* $d\mu_\ell$ *is not of the form considered previously, since the atom* N *has a prescribed position.*

2.3 Numerical tests

For our numerical tests, we choose the pair interaction potential

$$W(x) = \frac{1}{2}(x-1)^4 + \frac{1}{2}x^2 \tag{2.37}$$

shown on Figure 2.4. Note that $W(x)$ grows fast enough to $+\infty$ when $|x| \to +\infty$, such that assumptions (2.8) and (2.20) are satisfied. Note also that we have made no assumption on the convexity of W in Theorems 1 and 2. We consider here a convex potential. At the end of this section, we will consider a non-convex example (see (2.38)), and show that we obtain similar conclusions.

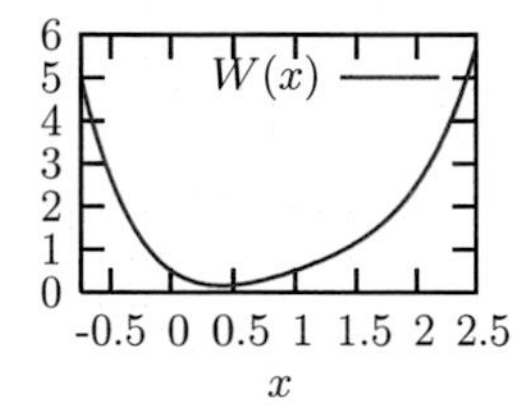

Figure 2.4: The potential W (defined by (2.37)) chosen for the tests.

We first consider the computation of ensemble averages, and we again restrict ourselves to the case of two repatoms $u^0 = 0$ and u^N. This is just for simplicity and for the sake of demonstrating the feasibility and the interest of our approach. The case of N_r repatoms may be treated likewise.

We choose an observable $A(x)$, and we compare the following four quantities:

(i) the exact average $\langle A \rangle_N$ defined by (2.2). Following (1.3)-(1.4), this quantity is computed as the long-time average of $A(u^N(t))$ along the full system dynamics

$$du_t = -\nabla_u E(u_t)\, dt + \sqrt{2/\beta}\, dB_t \quad \text{in } \mathbb{R}^N.$$

This equation is numerically integrated with the forward Euler scheme (also called the Euler-Maruyama scheme), with a small time step Δt:

$$u_{n+1} = u_n - \Delta t\, \nabla_u E(u_n) + \sqrt{2\Delta t/\beta}\, G_n$$

where G_n is a N-dimensional vector of random variables distributed according to a Gaussian normal law. In practice, we have simulated many independent realizations of this SDE, in order to compute error bars for $\langle A \rangle_N$.

(ii) a harmonic type approximation of $\langle A \rangle_N$, based on the 'interpolation + harmonic expansion' procedure outlined above. That is, we introduce $E_{\rm HA}$ defined by (2.32), and we approximate $\langle A \rangle_N$ by

$$\langle A \rangle_N^{\rm HA} := \frac{\displaystyle\int_{\mathbb{R}} A(x)\ \exp\left[-\beta E_{\rm HA}(x)\right]\, dx}{\displaystyle\int_{\mathbb{R}} \exp\left[-\beta E_{\rm HA}(x)\right]\, dx}.$$

(iii) a Law of Large Numbers (LLN) type approximation of $\langle A \rangle_N$, which consists in approximating $\langle A \rangle_N$ by $A(y^*)$, following Theorem 1.

(iv) a refined approximation, which consists in approximating $\langle A \rangle_N$ by $\langle A \rangle_N^{\rm approx}$ defined by (2.12), following Theorem 1.

Note that only one-dimensional integrals are needed for approximations (ii), (iii) and (iv). They can be computed with a high accuracy.

We plot on Figure 2.5 these four quantities, for increasing values of N (the temperature is fixed at $1/\beta = 1$), for the observable $A(x) = \exp(x)$. On Figure 2.6, we compare the same quantities, now as functions of the temperature, for $N = 100$ and for $N = 10$. We here work with $A(x) = x^2$, for which $\langle A \rangle_N = \langle A \rangle_N^{\rm approx}$.

As expected, the thermodynamic limit strategies (iii) and (iv) better agree with the full atom calculation, whatever the temperature, provided the number of eliminated atoms is large (note that the strategy (iv) is very accurate even for the small value $N = 10$, at the temperature $1/\beta = 1$). Approximation (ii) is clearly ineffective for high temperatures. On the other hand, for a sufficiently small temperature and a sufficiently small number of eliminated atoms, this approximation is close to the full atom result. However, even for the small values $N = 10$ and $1/\beta = 0.2$, our asymptotic result $\langle A(u^N) \rangle_N^{\rm approx} = 1.6299$ (for $A(x) = \exp(x)$) is closer to the exact result $\langle A(u^N) \rangle_N = 1.6303 \pm 0.0008$ than the harmonic approximation result $\langle A(u^N) \rangle_N^{\rm HA} = 1.6469$.

Remark 10:
As in Remark 3, we emphasize that the computations reported on here do not account for constraints on the positions of atoms. Analogous computations, that account for constraints, may be performed, using the potential W_c defined by (2.6)-(2.37)*. They provide similar conclusions, as can be seen on Figure 2.7, which is very similar to Figure 2.6.*

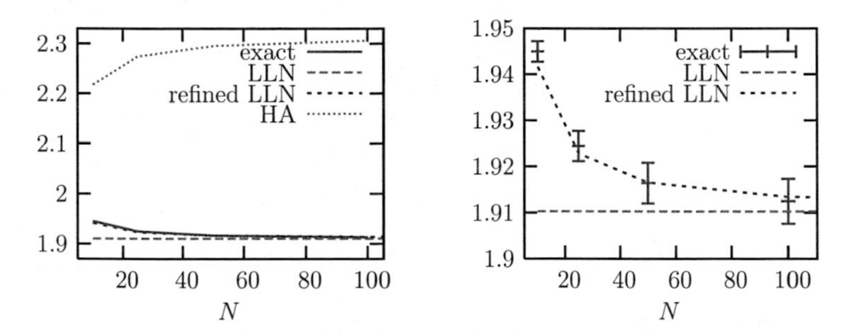

Figure 2.5: Convergence, as N increases, of $\langle A(u^N)\rangle_N$ (exact), of $\langle A(u^N)\rangle_N^{\text{approx}}$ (refined LLN) and of $\langle A(u^N)\rangle_N^{\text{HA}}$ (HA) and comparison to $A(y^*)$ (LLN) (temperature $1/\beta = 1$, observable $A(x) = \exp(x)$; we have performed computations for $N = 10$, 25, 50 and 100; on the right graph, we only plot the most accurate results with error bars).

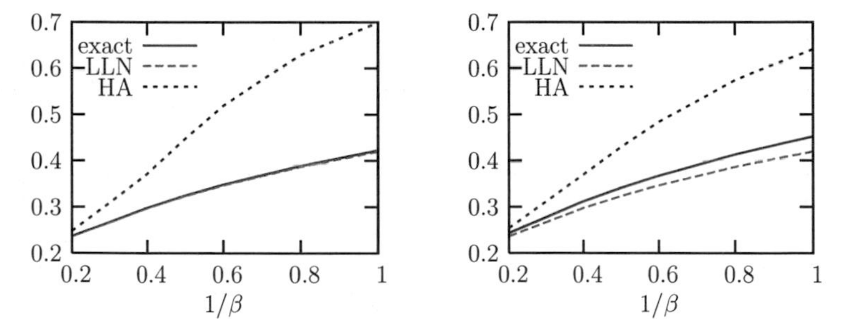

Figure 2.6: We plot $\langle A(u^N)\rangle_N = \langle A(u^N)\rangle_N^{\text{approx}}$ (exact), $\langle A(u^N)\rangle_N^{\text{HA}}$ (HA) and $A(y^*)$ (LLN) as functions of the temperature $1/\beta$: on the left graph, $N = 100$; on the right graph, $N = 10$ (observable $A(x) = x^2$).

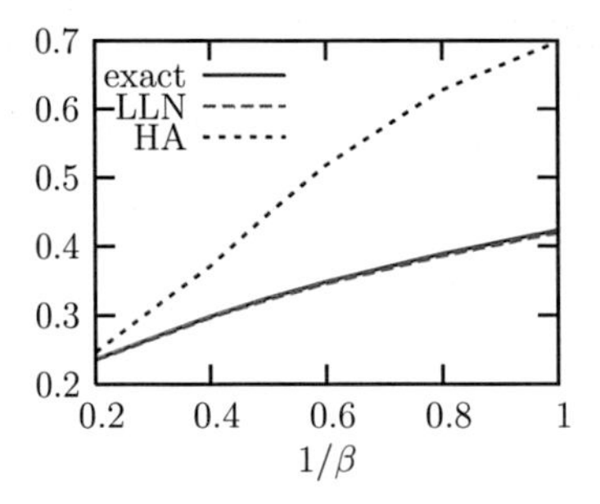

Figure 2.7: We plot $\langle A(u^N)\rangle_N = \langle A(u^N)\rangle_N^{\text{approx}}$ (exact), $\langle A(u^N)\rangle_N^{\text{HA}}$ (HA) and $A(y^*)$ (LLN) as functions of the temperature $1/\beta$ ($N = 100$, observable $A(x) = x^2$). The potential energy is of type (2.6): it is equal to $W(x)$ defined by (2.37) if $x > 0$, and $+\infty$ otherwise.

We now turn to the case of the free energy F_N. We are going to compare different approximations of its derivative. The full atom value $F'_N(x)$ is computed as the ensemble average (2.36). We compare this quantity with

(i) its large N limit $F'_\infty(x)$, where F_∞ is defined by (2.22), on the one hand,

(ii) and, on the other hand, its harmonic type approximation $F'_{\text{HA}}(x)$, where F_{HA} is defined by (2.33). We have

$$F'_{\text{HA}}(x) = W'(x) + \frac{1}{2\beta} \frac{W'''(x)}{W''(x)}.$$

We briefly detail how we compute $F'_\infty(x)$. Let ξ_x be the unique real number at which the supremum in (2.22) is attained. We have

$$F'_\infty(x) = \frac{\xi_x}{\beta}.$$

The Euler-Lagrange equation solved by ξ_x is (2.28), that we recast as

$$z^{-1} \int_{\mathbb{R}} (x - y) \exp(\xi_x y) \exp(-\beta W(y))\, dy = 0.$$

Let us introduce $G(y, \xi) = (x - y) \exp(\xi y)$. We hence look for ξ_x such that $\mathbb{E}_\mu [G(Y, \xi_x)] = 0$, where the scalar random variable Y is distributed according to the probability measure $\mu(y) = z^{-1} \exp(-\beta W(y))$. The Robbins-Monroe algorithm [KC78] can be used to compute ξ_x, hence $F'_\infty(x)$.

We first study the convergence of $F'_N(x)$ to $F'_\infty(x)$ as N increases, for a fixed chain length $x = 1.4$ and a fixed temperature $1/\beta = 1$. Results are shown on Figure 2.8. We indeed observe that $F'_N(x) \to F'_\infty(x)$ when $N \to +\infty$.

We now compare the two approximations (i) and (ii) of $F'_N(x)$, for $N = 100$ and $1/\beta = 1$. Results are shown on Figure 2.9. We observe that $F'_\infty(x)$ is a very good approximation of $F'_N(x)$. As expected, the temperature is too high for the harmonic approximation to provide an accurate approximation of $F'_N(x)$.

On Figure 2.10, we plot $F'_\infty(x)$ for several temperatures, as well as its zero temperature limit, which is $W'(x)$ (see Remark 7).

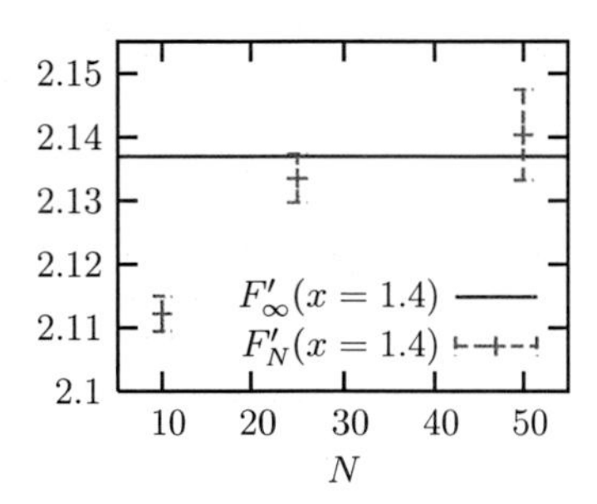

Figure 2.8: Convergence of $F'_N(x)$ (shown with error bars) to $F'_\infty(x)$ as N increases (temperature $1/\beta = 1$, fixed chain length $x = 1.4$).

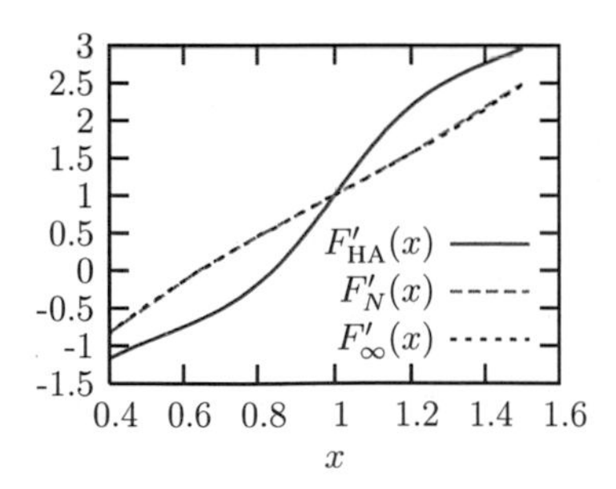

Figure 2.9: We plot $F'_N(x)$, $F'_\infty(x)$ and $F'_{\text{HA}}(x)$, for the temperature $1/\beta = 1$ and $N = 100$. On the scale of the figure, $F'_N(x)$ and $F'_\infty(x)$ are on top of each other.

Up to here, we have used the convex potential (2.37). For the sake of completeness, we now briefly consider the case of a non-convex potential W. We choose the toy-model

$$W(x) = (x^2 - 1)^2, \tag{2.38}$$

which corresponds to a double-well potential. We first study the convergence of $F'_N(x)$ to $F'_\infty(x)$ as N increases, for a fixed chain length $x = 0.5$ and a fixed temperature $1/\beta = 1$. Results are shown on Figure 2.11. As for the convex potential case, we observe that $F'_N(x) \to F'_\infty(x)$ when $N \to +\infty$. On Figure 2.12, we plot $F'_\infty(x)$ for several temperatures. Numerical results are consistent with the small temperature limit $\lim_{T\to 0} F'_\infty(x) = (W^*)'(x)$, where W^* is the convex envelop of W.

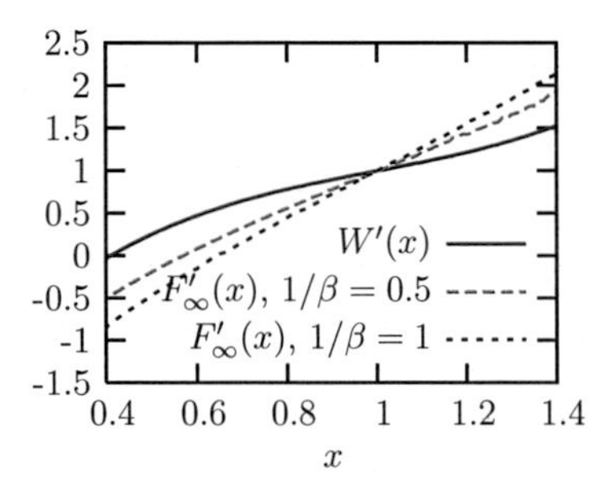

Figure 2.10: $F'_\infty(x)$ for different temperatures.

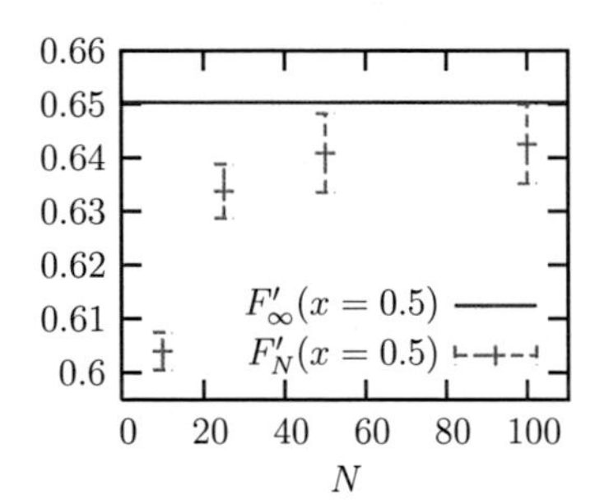

Figure 2.11: Convergence of $F'_N(x)$ (shown with error bars) to $F'_\infty(x)$ as N increases (temperature $1/\beta = 1$, fixed chain length $x = 0.5$), in the case of the double-well potential (2.38).

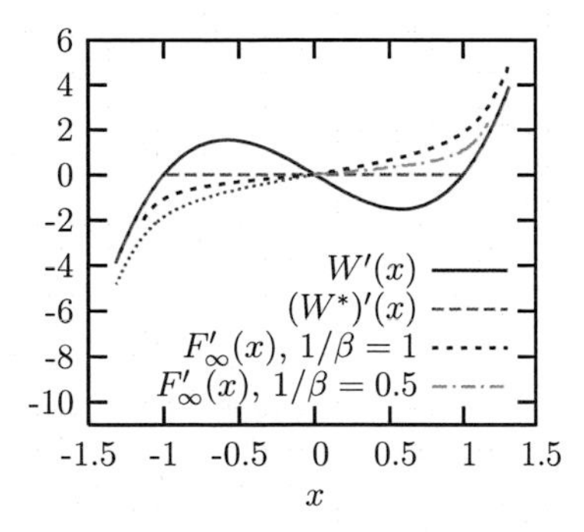

Figure 2.12: $F'_\infty(x)$ for different temperatures, in the case of the double-well potential (2.38).

Note that, in view of its definition (2.22), F_∞ is always a convex function. Hence, as in the zero temperature case, we observe, in this one-dimensional setting, that the macroscopic constitutive law $\ell \mapsto F_\infty(\ell)$ is a convex function.

3 The NNN case and some extensions

In this Section, we first consider the case of a NNN interacting system. The analysis is detailed in Section 3.1. In Section 3.2, we point out some possible extensions, first the NNNN case (still for one-dimensional systems) and second the case of linear polymer chains, where atoms sample the physical space $\mathbb{R}^3$.

3.1 The next-to-nearest neighbor (NNN) case

We now consider the next-to-nearest neighbour case. It turns out that, for the computation of ensemble averages as well as for other questions, this case is significantly more intricate than the NN case. Our strategy, based on the Law of Large Numbers, will be similar to that used for the NN case, but the object manipulated are not independent random variables any longer. Markov chains are the right notion formalizing the situation mathematically.

We begin by introducing the rescaled atomistic energy, similarly to (2.1):

$$E\left(u^1, \ldots, u^N\right) = \sum_{i=1}^{N} W_1\left(\frac{u^i - u^{i-1}}{h}\right) + \sum_{i=1}^{N-1} W_2\left(\frac{u^{i+1} - u^{i-1}}{h}\right).$$

As above, we set $u^0 = 0$, and we introduce the change of variables (2.3), replacing $(u^i - u^{i-1})/h$ by the inter-atomic distances y_i. Recall from (2.4) that

$u^N = \frac{1}{N}\sum_{i=1}^N y_i$. The ensemble average $\langle A\rangle_N$ of an observable that depends only on the right-end atom therefore writes

$$\begin{aligned}\langle A\rangle_N &= Z^{-1}\int_{\mathbb{R}^N} A\left(u^N\right)\exp\left(-\beta E\left(u^1,\ldots,u^N\right)\right)du^1\ldots du^N \\ &= Z^{-1}\int_{\mathbb{R}^N} A\left(\frac{1}{N}\sum_{i=1}^N y_i\right) e^{-\beta\left(\sum_{i=1}^N W_1(y_i)+\sum_{i=1}^{N-1}W_2(y_i+y_{i+1})\right)}\,dy_1\ldots dy_N.\end{aligned} \tag{3.1}$$

The key ingredient is now to see the above expression, as N goes to infinity, as an asymptotics for a *discrete-time Markov chain.* The asymptotics of Markov chains being a mathematical problem much more involved than that of i.i.d. sequences, we restrict ourselves to the computation of the average of an observable. The asymptotic behavior of the free energy may be studied, applying a Large Deviations Principle for Markov chains (see for instance [dH00, Theorem IV.3]). We will not pursue in this direction.

Section 3.1.1 deals with the case of two repatoms (namely $u^0 = 0$ and u^N), while Section 3.1.2 indicates the changes in order to deal with more than two repatoms. Numerical results will be reported in Section 3.1.3.

3.1.1 Limit of the average, the case of two repatoms

In order to compute $\lim_{N\to+\infty}\langle A\rangle_N$, we introduce the notation

$$f(x,y) := \exp(-\beta W_2(x+y))\ \exp(-\beta W_1(y)).$$

Equation (3.1) rewrites

$$\begin{aligned}\langle A\rangle_N = Z^{-1}\int_{\mathbb{R}^N} A\left(\frac{1}{N}\sum_{i=1}^N y_i\right) e^{-\beta W_1(y_1)}\ f(y_1,y_2)\ldots \\ \ldots f(y_{N-1},y_N)\,dy_1\,\ldots\,dy_N.\end{aligned} \tag{3.2}$$

Our method consists in considering the sequence of variables $(y_1,\ldots,y_N)$ in (3.2) as a realization of a Markov chain with kernel $f(\cdot,\cdot)$. However, the slight technical difficulty at this stage is that the kernel f is not normalized, since in general

$$\int_{\mathbb{R}} f(y_1,y_2)\,dy_2 = \int_{\mathbb{R}}\exp(-\beta W_2(y_1+y_2))\ \exp(-\beta W_1(y_2))\,dy_2 \neq 1.$$

A standard trick of Probability theory allows to circumvent this difficulty. Introduce

$$\overline{f}(x,y) := \exp\left[-\beta W_2(x+y) - \frac{\beta}{2}W_1(x) - \frac{\beta}{2}W_1(y)\right].$$

Note that $\overline{f}$ is a symmetric function (whereas f is not), hence the operator

$$P\phi(y) = \int_{\mathbb{R}}\overline{f}(y,z)\,\phi(z)\,dz \tag{3.3}$$

is self-adjoint on $L^2(\mathbb{R})$. Consider then the variational problem

$$\max_{\psi \in L^2(\mathbb{R})} \left\{ \int_{\mathbb{R}^2} \psi(y)\, \psi(z)\, \overline{f}(y,z)\, dy\, dz; \ \int_{\mathbb{R}} \psi^2(y)\, dy = 1 \right\}. \tag{3.4}$$

Using standard tools of spectral theory of self-adjoint compact operators [DS63], it is possible to prove that this problem has a maximizer (we denote it by ψ_1), and that, up to changing ψ_1 into $-\psi_1$, the maximizer is unique. In addition, one can choose ψ_1 such that $\psi_1 > 0$. The Euler-Lagrange equation of (3.4) reads

$$\lambda\, \psi_1(y) = \int_{\mathbb{R}} \overline{f}(y,z)\, \psi_1(z)\, dz$$

for some λ, and we recognize an eigenvalue problem, $P\psi_1 = \lambda\, \psi_1$, for the operator (3.3). Multiplying the above equation by $\psi_1(y)$ and integrating, we obtain

$$\lambda = \int_{\mathbb{R}^2} \psi_1(y)\, \psi_1(z)\, \overline{f}(y,z)\, dy\, dz > 0.$$

We now define

$$g(y,z) := \frac{\psi_1(z)}{\lambda \psi_1(y)}\, \overline{f}(y,z). \tag{3.5}$$

By construction,

$$\int_{\mathbb{R}} g(y,z)\, dz = 1, \quad \int_{\mathbb{R}} \psi_1^2(y)\, g(y,z)\, dy = \psi_1^2(z).$$

The average (3.2) now reads

$$\langle A \rangle_N = Z_g^{-1} \int_{\mathbb{R}^N} A\left(\frac{1}{N} \sum_{i=1}^N y_i \right) \psi_1(y_1)\, e^{-\frac{\beta}{2} W_1(y_1)} \times g(y_1, y_2) \dots g(y_{N-1}, y_N)\, \frac{e^{-\frac{\beta}{2} W_1(y_N)}}{\psi_1(y_N)}\, dy_1 \dots dy_N,$$

with $Z_g = \int_{\mathbb{R}^N} \psi_1(y_1)\, e^{-\frac{\beta}{2} W_1(y_1)}\, g(y_1, y_2) \dots g(y_{N-1}, y_N)\, \frac{e^{-\frac{\beta}{2} W_1(y_N)}}{\psi_1(y_N)}\, dy_1 \dots dy_N$, and where $(y_1, \dots, y_N)$ may now be seen as a realization of a *normalized* Markov chain of kernel g, with invariant probability measure ψ_1^2. We assume that $\overline{f}$ decays fast enough at infinity (which ensures for instance that (3.4) is well-posed) and that it is positive. This latter assumption ensures that the Markov chain satisfies the following accessibility condition: for any $x \in \mathbb{R}$, and for any measurable set $\mathcal{B} \subset \mathbb{R}$ of positive Lebesgue measure, we have $\int_{\mathcal{B}} g(x,y)\, dy > 0$. Under this property, combined with the existence of an invariant probability measure ($\psi_1^2(y)\, dy$ in the present case), it is known (see [MT09, Theorem 17.1.7]) that the invariant measure is unique, and that the Markov chain satisfies a Law of Large Numbers with respect to it. We now state a direct corollary of this general result that applies to our context.

Theorem 3:
Assume that A is continuous, and satisfies the following conditions:

$$\exists p \geq 0, \quad \exists C > 0, \quad \forall x \in \mathbb{R}, \quad |A(x)| \leq C(1+|x|^p).$$

Under the assumptions that $W_1, W_2 \in L^1_{\text{loc}}(\mathbb{R})$ are bounded from below, that $e^{-\beta W_1}, e^{-\beta W_2} \in W^{1,1}_{\text{loc}}(\mathbb{R})$ and that

$$\forall q \geq 0, \int_{\mathbb{R}} |x|^q \; e^{-\beta W_1(x)} dx < +\infty \quad \text{and} \quad \int_{\mathbb{R}} |x|^q \; e^{-\beta W_2(x)} dx < +\infty, \tag{3.6}$$

the ergodic theorem for Markov chains [MT09] yields

$$\lim_{N \to +\infty} \langle A \rangle_N = A(y^*) \quad \text{where} \quad y^* := \int_{\mathbb{R}} y \; \psi_1^2(y) \, dy.$$

Remark 11:
Note that, for the result to hold true, (3.6) is not needed. The existence of the moment of order p is sufficient. However, assumption (3.6) will be useful for Theorem 4 below.

Remark 12:
It might sound a little strange that ψ_1 is the eigenvector of the transition operator P defined by (3.3), whereas the invariant measure of the chain is ψ_1^2. This is explained by the following fact: the expectation value of y_i is equal to

$$\mathbb{E}(y_i) = \int_{\mathbb{R}} x \; (P^{i-1}\varphi_0)(x) \left(P^{N-i}\varphi_1\right)(x) \, dx,$$

for some initial laws φ_0 and φ_1, where P^i is recursively defined by $P^i\varphi = P[P^{i-1}\varphi]$ (recall that P is the operator defined by (3.3)). Hence, if $1 \ll i \ll N$, then both $P^{i-1}\varphi_0$ and $P^{N-i}\varphi_1$ converge to the eigenvector of P associated with the largest eigenvalue, that is, ψ_1 (hence the appearance of ψ_1^2 rather than ψ_1). This is explained in more details in Section 3.2.1 below in the case of a non self-adjoint transition operator P.

Here again, it is possible to compute the next terms in the expansion of $\langle A \rangle_N$ in powers of $1/N$. However, the computations are much more intricate than in the i.i.d. case. The terms of the expansion here contain covariance terms, together with terms containing the initial state of the Markov chain. As an example, we give the first term of the expansion in the following theorem.

Theorem 4:
Assume that A is of class C^3. Then, under the assumptions of Theorem 3, we have

$$\langle A \rangle_N = A(y^*) + \frac{1}{N} A'(y^*) \sum_{i \geq 1} \mathbb{E}\left(Y_i - y^*\right) + \frac{\sigma^2}{2N} A''(y^*) + O\left(\frac{1}{N^2}\right), \tag{3.7}$$

with

$$\sigma^2 = \int_{\mathbb{R}} (x - y^*)^2 \ \psi_1^2(x)\, dx + 2 \sum_{i \geq 2} \mathbb{E}\left((\widetilde{Y}_i - y^*)(\widetilde{Y}_1 - y^*) \right), \tag{3.8}$$

where $(Y_i)_{i\geq 1}$ *and* $\left(\widetilde{Y}_i\right)_{i\geq 1}$ *are Markov chains of initial law* $Z_1^{-1}\psi_1 e^{-\frac{\beta}{2}W_1}$ *and* ψ_1^2 *respectively, and of transition kernel* g. *Moreover, the series appearing in* (3.7) *and* (3.8) *converge exponentially fast.*

Remark 13:
Let us mention that σ^2 *defined by* (3.8) *is exactly the variance appearing in the Central Limit Theorem for Markov chains [MT09, Theorem 17.0.1].*

In addition, we see that in the special case of i.i.d. random variables, the second term of (3.8) *vanishes, together with the term proportional to* $A'(y^*)$ *in* (3.7). *We then recover estimate (2.11)-(2.12).*

Remark 14:
The assumptions of Theorems 3 and 4 are not sharp. However, they allow for simple proofs, and for a wide variety of interaction potentials W_1 *and* W_2.

Remark 15:
Again, as in Remark 3, constraints on the positions of the atoms may be accounted for.

Note that Theorem 4 suggests a strategy for numerically computing the terms of (3.7). Indeed, it is possible to compute numerically ψ_1 by discretizing (3.4). Numerical integration then allows to compute y^* and the variance $\int_{\mathbb{R}}(x-y^*)^2\psi_1^2(x)dx$. The computation of the infinite sums in (3.7)-(3.8) is then performed using a simulation of the corresponding Markov chains and taking the expectation value. Note that the law of Y_i converges exponentially fast (when $i \to \infty$) to the invariant measure ψ_1^2 due to the existence of a spectral gap for the transition operator. Hence, the terms $\mathbb{E}\,(Y_i - y^*)$ and $\mathbb{E}\left((\widetilde{Y}_i - y^*)(\widetilde{Y}_1 - y^*)\right)$ that appear in both sums decay exponentially fast, and only a few terms are needed in practice. We will observe in Section 3.1.3 that, on our test example, $A(y^*)$ is already a good approximation of $\langle A \rangle_N$. Hence, we have not implemented the strategy just described.

3.1.2 More than two repatoms

We explain in this Section how the results of Section 3.1.1 can be adapted to the case when more than two repatoms are considered.

We thus consider the following setting: we have $N + M + 1$ atoms of positions u^i, $0 \leq i \leq M + N$, and u^0, u^N and u^{N+M} are the repatoms. We assume that N and M are such that $N/(N + M) = L_1$ is fixed. To remove translation invariance, we set $u^0 = 0$. Since the atoms on the right of u^N will not play the same role as those on the left, we denote their distance differently:

$$y_i := \frac{u^i - u^{i-1}}{h} \quad \forall 1 \le i \le N, \quad z_i := \frac{u^{i+1+N} - u^{i+N}}{h} \quad \forall 0 \le i \le M-1,$$

where $h = 1/(N+M)$. We assume that the observable A is a function of $u^N - u^0$ and $u^{N+M} - u^N$. Hence, the expectation value of A reads:

$$\begin{aligned} \langle A \rangle_{N,M} &= Z^{-1} \int_{\mathbb{R}^{N+M}} A\left(\frac{1}{N+M}\sum_{i=1}^{N} y_i, \frac{1}{N+M}\sum_{i=0}^{M-1} z_i\right) e^{-\frac{\beta}{2}W_1(y_1)} \\ &\prod_{i=1}^{N-1} \overline{f}(y_i, y_{i+1})\, e^{-\frac{\beta}{2}W_1(y_N)}\, e^{-\beta W_2(y_N+z_0)}\, e^{-\frac{\beta}{2}W_1(z_0)} \\ &\prod_{i=0}^{M-2} \overline{f}(z_i, z_{i+1})\, e^{-\frac{\beta}{2}W_1(z_{M-1})}\, dy\, dz, \end{aligned} \tag{3.9}$$

where, as before, we have set $\overline{f}(x,y) = \exp\left[-\beta W_2(x+y) - \frac{\beta}{2}W_1(x) - \frac{\beta}{2}W_1(y)\right]$. Here again, we may use ψ_1 that solves (3.4) in order to rewrite (3.9) as the expectation value of a function of two independent Markov chains. Indeed, ψ_1 and g being defined as before (see (3.4) and (3.5)), we have

$$\begin{aligned} \langle A \rangle_{N,M} &= Z^{-1} \int_{\mathbb{R}^{N+M}} A\left(\frac{1}{N+M}\sum_{i=1}^{N} y_i, \frac{1}{N+M}\sum_{i=0}^{M-1} z_i\right) \psi_1(y_1)\, e^{-\frac{\beta}{2}W_1(y_1)} \\ &\prod_{i=1}^{N-1} g(y_i, y_{i+1})\, \frac{e^{-\frac{\beta}{2}W_1(y_N)}}{\psi_1(y_N)}\, e^{-\beta W_2(y_N+z_0)}\, \psi_1(z_0)\, e^{-\frac{\beta}{2}W_1(z_0)} \\ &\prod_{i=0}^{M-2} g(z_i, z_{i+1})\, \frac{e^{-\frac{\beta}{2}W_1(z_{M-1})}}{\psi_1(z_{M-1})}\, dy\, dz, \\ &= \mathbb{E}\left[A\left(\frac{1}{N+M}\sum_{i=1}^{N} Y_i, \frac{1}{N+M}\sum_{i=0}^{M-1} Z_i\right) \frac{e^{-\frac{\beta}{2}W_1(Y_N) - \frac{\beta}{2}W_1(Z_{M-1}) - \beta W_2(Y_N+Z_0)}}{\psi_1(Y_N)\psi_1(Z_{M-1})}\right] \\ &\times \left(\mathbb{E}\left[\frac{e^{-\frac{\beta}{2}W_1(Y_N) - \frac{\beta}{2}W_1(Z_{M-1}) - \beta W_2(Y_N+Z_0)}}{\psi_1(Y_N)\psi_1(Z_{M-1})}\right]\right)^{-1}, \end{aligned} \tag{3.10}$$

where the sequences $(Y_i)_{i\ge1}$ and $(Z_i)_{i\ge0}$ are two independent realizations of a Markov chain of initial law $\psi_1 e^{-\frac{\beta}{2}W_1}$, and of transition kernel g. These Markov chains have exactly the same properties as the chain of Section 3.1.1. Hence, we may use again the ergodic theorem as before to prove that:

$$\frac{1}{N+M}\sum_{i=1}^{N} Y_i \longrightarrow L_1 y^*, \quad \frac{1}{N+M}\sum_{i=0}^{M-1} Z_i \longrightarrow L_2 y^*,$$

almost surely, with $L_1 = N/(N+M)$ and $L_2 = M/(N+M)$. Thus, the expectation values in (3.10) simplify, allowing to prove:

Theorem 5:
Assume that A, W_1 and W_2 satisfy the assumptions of Theorem 3. Assume in addition that $L_1 = \frac{N}{N+M}$ is fixed, and set $L_2 = \frac{M}{N+M} = 1 - L_1$. Then, we have

$$\lim_{N,M\to+\infty} \langle A\rangle_{N,M} = A\left(L_1 y^*, L_2 y^*\right) \quad \textit{with} \quad y^* = \int_{\mathbb{R}} y\, \psi_1^2(y)\, dy.$$

Here again, it is possible to use an expansion in powers of $1/N$ and $1/M$ of the expectation value $\langle A\rangle_{N,M}$. For simplicity, we restrict ourselves to the expansion at first order, and consider the case $N = M$. We assume that the hypotheses of Theorem 4 and Theorem 5 are satisfied. We then have

$$\begin{aligned}\langle A\rangle_{N,N} &= A\left(L_1 y^*, L_2 y^*\right) + \frac{\sigma^2}{2N}\Delta A(L_1 y^*, L_2 y^*) \\ &\quad + \frac{1}{N}\left[\partial_1 A(L_1 y^*, L_2 y^*) + \partial_2 A(L_1 y^*, L_2 y^*)\right] \sum_{i\geq 1} \mathbb{E}\left(Y_i - y^*\right) + O\left(\frac{1}{N^2}\right),\end{aligned}$$

with $L_1 = L_2 = 1/2$ and where σ is defined by (3.8). We have not implemented this formula, since, on our test example, $A\left(L_1 y^*, L_2 y^*\right)$ is already a good approximation of $\langle A\rangle_{N,N}$.

3.1.3 Numerical results

For the NNN model, we choose the potentials

$$W_1(x) = \frac{1}{2}(x-1)^4 + \frac{1}{2}x^2 \quad \text{and} \quad W_2(x) = \frac{1}{4}(x-2.1)^4.$$

Note that other choices are possible, such as $W_1 \equiv W_2$, or $W_2(x) = W_1(x/2)$ (such that the equilibrium distances of W_1 and W_2 are compatible). We have chosen W_2 such that we observe a significant dependence of ensemble averages (for instance of the mean length $\langle u^N\rangle_N$ of the chain) with respect to temperature.

It is important that W_1 and W_2 grow fast enough at infinity, such that assumptions of Theorem 3 are satisfied (in particular assumption (3.6)). As in the NN case, we do not need any convexity assumption on W_1 and W_2.

We consider two cases:

- the chain consists of $N+1$ atoms, there are two repatoms $u^0 = 0$ and u^N, and the observable only depends on the right end atom u^N. We aim at computing $\langle A(u^N)\rangle_N$. This is the situation of Section 3.1.1.
- the chain consists of $2N+1$ atoms, there are three repatoms $u^0 = 0$, u^N and u^{2N}, and the observable depends on u^N and $u^{2N} - u^N$. We aim at computing $\langle A(u^N, u^{2N} - u^N)\rangle_{2N}$. This is a situation covered by Section 3.1.2.

Theorems 3 and 5 respectively provide the asymptotics $\lim_{N\to+\infty}\langle A(u^N)\rangle_N = A(y^*)$ and $\lim_{N\to+\infty}\langle A(u^N, u^{2N} - u^N)\rangle_{2N} = A\left(\frac{1}{2}y^*, \frac{1}{2}y^*\right)$.

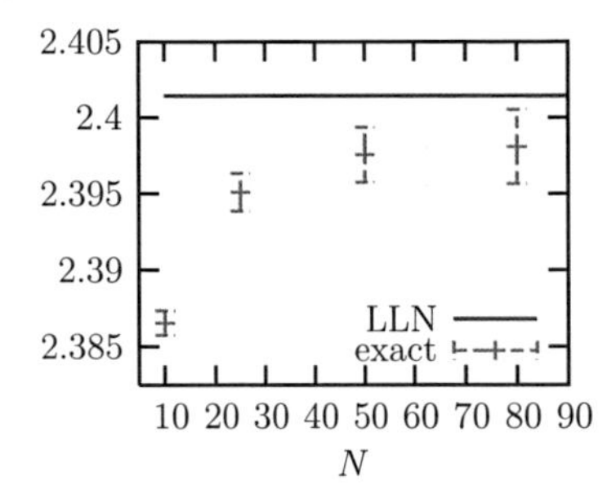

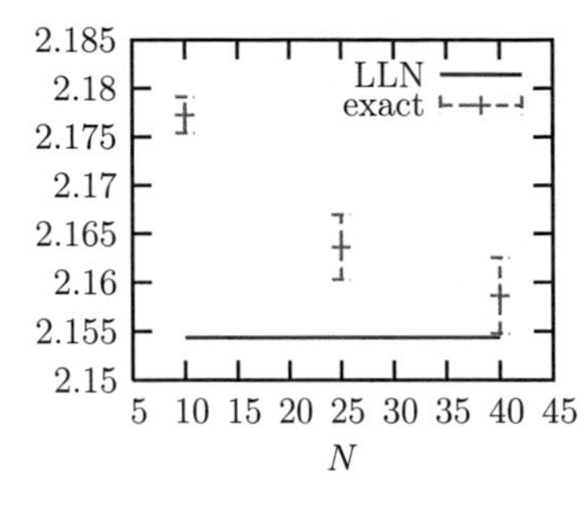

Figure 3.1: Left graph: convergence, as N increases, of $\langle A(u^N)\rangle_N$ (exact) to $A(y^*)$ (LLN), at the temperature $1/\beta = 1$, for $A(x) = \exp(x)$. Right graph: convergence, as N increases, of $\langle A(u^N, u^{2N} - u^N)\rangle_{2N}$ (exact) to $A(y^*/2, y^*/2)$ (LLN), at the temperature $1/\beta = 1$, for $A(x,y) = \exp(2x(x+y))$.

We first study the convergence of ensemble averages at the temperature $1/\beta = 1$, as N increases. Results are shown on Figure 3.1, for a particular choice of observable (we have performed the same tests with other observables, with similar conclusions). We indeed observe that the ensemble averages of the full atom system converge to their Law of Large Numbers (LLN) limit, in both cases of two and three repatoms. Note that the exact result for $N = 10$ is already very well approximated by the asymptotic limit, namely $A(y^*)$ in the two repatoms case, $A(y^*/2, y^*/2)$ in the three repatoms case.

We next study the averages as functions of the temperature, for $N = 100$. Results are shown on Figure 3.2. We observe an excellent agreement between the full atom value and the asymptotic limit, in both cases of two and three repatoms, whatever the temperature.

3.2 Extensions

In this Section, we briefly explain that our strategy carries out to more general cases.

3.2.1 The NNNN case

The case of any finite range interaction may be treated in the same way as we treated the NNN case in Section 3.1.1. Indeed, consider for instance the case of next to next to nearest neighbor interaction (NNNN). In such a case, we are lead

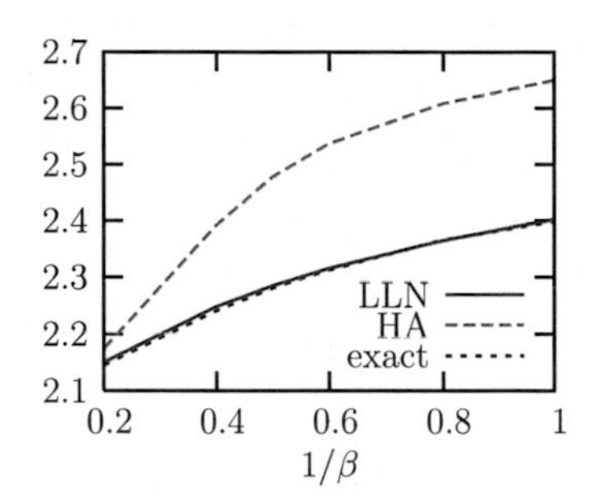

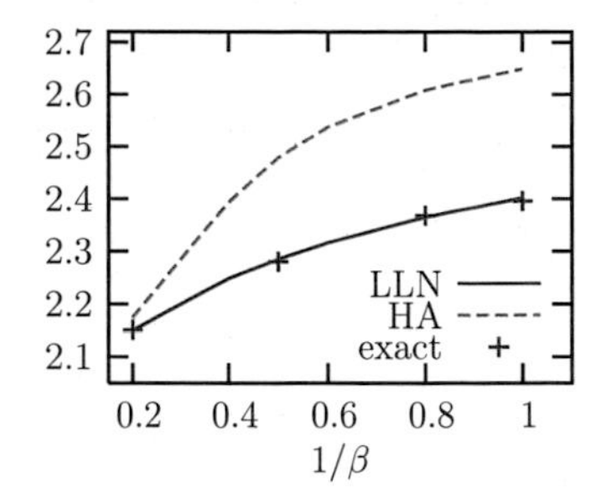

Figure 3.2: Left graph: we plot $\langle A(u^N)\rangle_N$ (exact), $\langle A(u^N)\rangle_N^{\mathrm{HA}}$ (HA) and $A(y^*)$ (LLN) as functions of the temperature $1/\beta$ ($N = 100$, $A(x) = \exp(x)$). Right graph: we plot $\langle A(u^N, u^{2N} - u^N)\rangle_{2N}$ (exact), $\langle A(u^N, u^{2N} - u^N)\rangle_{2N}^{\mathrm{HA}}$ (HA) and $A(y^*/2, y^*/2)$ (LLN) as functions of the temperature $1/\beta$ ($N = 100$, $A(x, y) = \exp(2x)$).

to consider (we go back here to the case of 2 repatoms for the sake of clarity):

$$\begin{aligned}\langle A\rangle_N &= Z^{-1}\int_{\mathbb{R}^N} A\left(\frac{1}{N}\sum_{i=1}^N y_i\right)\exp\Bigg[-\beta\sum_{i=1}^N W_1(y_i)\\ &\qquad -\beta\sum_{i=1}^{N-1} W_2(y_i + y_{i+1}) - \beta\sum_{i=1}^{N-2} W_3(y_i + y_{i+1} + y_{i+2})\Bigg]\, dy_1 \dots dy_N\\ &= Z^{-1}\int_{\mathbb{R}^N} A\left(\frac{1}{N}\sum_{i=1}^N y_i\right) b(y_{N-1}, y_N)\prod_{i=1}^{N-2} f(y_i, y_{i+1}, y_{i+2})\, dy_1 \dots dy_N,\end{aligned}$$

where we have set $f(x, y, z) = \exp\left[-\beta W_1(x) - \beta W_2(x+y) - \beta W_3(x+y+z)\right]$, and used the notation $b(y_{N-1}, y_N) = \exp\left[-\beta W_1(y_{N-1}) - \beta W_1(y_N) - \beta W_2(y_{N-1} + y_N)\right]$ for the boundary term. We assume, for the sake of simplicity, that N is even, i.e.

$$N = 2M,$$

and define the new variables

$$\xi_i = (y_{2i-1}, y_{2i}), \quad 1 \le i \le M.$$

Hence, we have

$$\langle A\rangle_N = Z^{-1}\int_{\mathbb{R}^{2M}} A\left(\frac{1}{2M}\sum_{i=1}^M \xi_i\cdot(1,1)\right) b(\xi_M)\prod_{i=1}^{M-1}\widetilde{f}[\xi_i, \xi_{i+1}]\, d\xi_1 \dots d\xi_M, \quad (3.11)$$

where

$$\widetilde{f}[(x, y), (z, t)] = f(x, y, z)\, f(y, z, t).$$

Hence, this change of variables allows to manipulate again a Markov chain, but in dimension 2. However, the renormalization trick we have used in the NNN case (see Section 3.1.1) cannot be used here, because it relies on the fact that the transition operator is self-adjoint. It is nevertheless possible to use the above structure in the following way: define the operator (on $L^2(\mathbb{R}^2)$)

$$[P\varphi](z,t) = \int_{\mathbb{R}^2} \widetilde{f}[(x,y),(z,t)]\ \varphi(x,y)\,dx\,dy,$$

together with its adjoint

$$[P^*\varphi](z,t) = \int_{\mathbb{R}^2} \widetilde{f}[(z,t),(x,y)]\ \varphi(x,y)\,dx\,dy.$$

Before studying the average (3.11), let us consider the average

$$\langle B \rangle = Z^{-1} \int_{\mathbb{R}^{2M}} B(\xi_i)\ b(\xi_M)\ \prod_{i=1}^{M-1} \widetilde{f}[\xi_i,\xi_{i+1}]\,d\xi_1 \dots d\xi_M$$

for a continuous and bounded function B. Then

$$\langle B \rangle = Z^{-1} \int_{\mathbb{R}^2} B(\xi)\ [P^{i-2}\varphi](\xi)\ [(P^*)^{M-i-1}\psi](\xi)\ d\xi, \tag{3.12}$$

where

$$\varphi(z,t) = \int_{\mathbb{R}^2} \widetilde{f}[(x,y),(z,t)]\,dx\,dy,$$

$$\psi(t,z) = \int_{\mathbb{R}^2} b(y,x)\ \widetilde{f}[(t,z),(y,x)]\,dx\,dy.$$

We assume that the operators P and P^* have a simple and isolated largest eigenvalue (which can be proved for many interactions, using for instance Krein-Rutman theorem [SW99]). Let us denote by ϕ and ϕ^* the corresponding eigenvectors in $L^2(\mathbb{R}^2)$, namely

$$P\phi = \lambda\phi, \quad P^*\phi^* = \lambda\phi^*,$$

where $\lambda = \sup\ \text{Spectrum}(P) = \sup\ \text{Spectrum}(P^*)$. If $1 \ll i \ll M$, we infer from (3.12) that

$$\langle B \rangle \longrightarrow Z_\infty^{-1} \int_{\mathbb{R}^2} B(\xi)\,\phi(\xi)\,\phi^*(\xi)\,d\xi,$$

where $Z_\infty = \int_{\mathbb{R}^2} \phi\,\phi^*$. This argument may be adapted to prove that the expectation value $\langle A \rangle_N$ defined by (3.11) converges:

$$\langle A \rangle_N \underset{N\to+\infty}{\longrightarrow} A\left(y^*\right),$$

where

$$y^* = \frac{\displaystyle\int_{\mathbb{R}^2} (\xi_1+\xi_2)\,\phi(\xi_1,\xi_2)\,\phi^*(\xi_1,\xi_2)\,d\xi_1\,d\xi_2}{\displaystyle\int_{\mathbb{R}^2} \phi(\xi_1,\xi_2)\,\phi^*(\xi_1,\xi_2)\,d\xi_1\,d\xi_2}.$$

3.2.2 Polymer chains

The considerations of Section 2 and Section 3.1 may be easily generalized to the case when the positions u^i of the atoms are not restricted to be in the real line, but are vectors of $\mathbb{R}^2$ or $\mathbb{R}^3$. The only important thing here is that they are indexed by a one dimensional parameter (here, $1 \leq i \leq N$). This is the case for instance if one considers a polymer chain (see Figure 3.3). In such a case, NN or NNN approximations are commonly used, in order to compute the average length of the chain (see [BCAH87] for instance). Our approximation strategy carries out to this case.

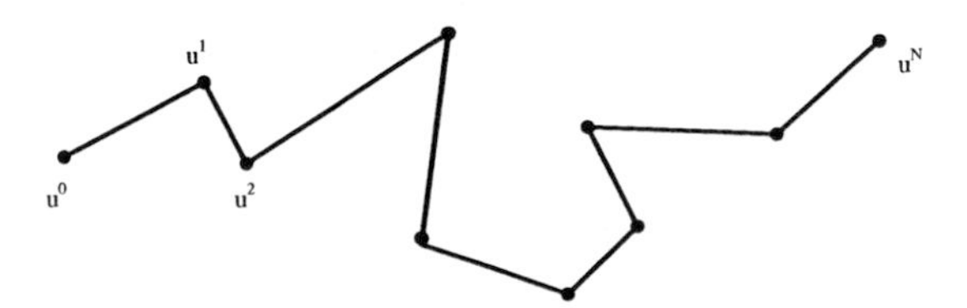

Figure 3.3: An example of polymer chain. The corresponding model is the same as the one studied in this article, except that the positions u^i are in $\mathbb{R}^3$.

Acknowledgments: Claude Le Bris and Frédéric Legoll would like to acknowledge stimulating discussions with Mitchell Luskin and Ellad Tadmor, University of Minnesota, Minneapolis. The hospitality of the Institute for Mathematics and its Applications (Minneapolis) and of the Weierstrass-Institut für Angewandte Analysis und Stochastik (Berlin) is gratefully acknowledged. Mathias Rousset provided some instrumental input from probability theory. This work has been initiated during the one-year stay (2005-2006) of the fourth author, Carsten Patz, at the Ecole Nationale des Ponts et Chaussées, as a visiting graduate student.

The research of Claude Le Bris and Frédéric Legoll is supported in part by the MEGAS non-thematic program (Agence Nationale de la Recherche, France) and by the INRIA, through the grant "Action de Recherche Collaborative" HYBRID.

References

[ANZEA90] M. Anitescu, D. Negrut, P. Zapol, A. El-Azab, *A note on the regularity of reduced models obtained by nonlocal quasi-continuum-like approach*, Mathematical Programming 118(2), pp 207–236, 2009.

[AG05] M. Arndt, M. Griebel, *Derivation of higher order gradient continuum models from atomistic models for crystalline solids*, SIAM J. Multiscale Model. Simul. 4(2), pp 531–562, 2005.

[AL05] M. Arndt, M. Luskin, *Error estimation and atomistic-continuum adaptivity for the quasicontinuum approximation of a Frenkel-Kantorova model*, SIAM J. Multiscale Model. Simul. 7(1), pp 147-170, 2008.

[BRR60] R. R. Bahadur, R. Ranga Rao, *On deviations of the sample mean*, Ann. Math. Statist. 31, pp 1015–1027, 1960.

[Bax82] R. J. Baxter, *Exactly solved models in statistical mechanics*, Academic Press, Inc. London, 1982.

[BO78] C. M. Bender, S. A. Orszag, *Advanced mathematical methods for scientists and engineers,* in: International Series in Pure and Applied Mathematics, McGraw-Hill, New York, 1978.

[BLBL05] X. Blanc, C. Le Bris, F. Legoll, *Analysis of a prototypical multiscale method coupling atomistic and continuum mechanics*, Math. Model. Numer. Anal. (M2AN) 39(4), pp 797-826, 2005.

[BLBL07a] X. Blanc, C. Le Bris, F. Legoll, *Analysis of a prototypical multiscale method coupling atomistic and continuum mechanics: the convex case*, Acta Math. Appl. Sinica 23(2), pp 209-216, 2007.

[BLBL02] X. Blanc, C. Le Bris, P.-L. Lions, *From molecular models to continuum mechanics*, Arch. Rat. Mech. Anal. 164, pp 341-381, 2002.

[BLBL07b] X. Blanc, C. Le Bris, P.-L. Lions, *Atomistic to continuum limits for computational materials science*, Math. Model. Numer. Anal. (M2AN) 41(2), pp 391–426, 2007.

[BCAH87] R. B. Bird, C. F. Curtiss, R. C. Armstrong, O. Hassager, *Dynamics of polymeric liquids. Volume 2: Kinetic theory.* Second edition, Wiley, New York, 1987.

[CLS07] E. Cancès, F. Legoll, G. Stoltz, *Theoretical and numerical comparison of some sampling methods for molecular dynamics*, Math. Model. Numer. Anal. (M2AN) 41(2), pp 351–389, 2007.

[CC02] S. Curtarolo, G. Ceder, *Dynamics of an inhomogeneously coarse grained multiscale system*, Phys. Rev. Lett. 88(25), 255504, 2002.

[DFP02] P. Deák, Th. Frauenheim, M. R. Pederson, Editors, *Computer simulation of materials at atomic level*, Wiley, 2000.

[DZ93] A. Dembo, O. Zeitouni, *Large deviations techniques and applications*, Jones and Bartlett Publishers, Boston, MA, 1993.

[Der07] B. Derrida, *Non-equilibrium steady states: fluctuations and large deviations of the density and of the current*, J. Stat. Mech. P07023, 2007.

[DL08] M. Dobson, M. Luskin, *Analysis of a force-based quasicontinuum approximation*, Math. Model. Numer. Anal. (M2AN) 42(1), pp 113–139, 2008.

[DL09a] M. Dobson, M. Luskin, *An analysis of the effect of ghost force oscillation on quasicontinuum error*, Math. Model. Numer. Anal. (M2AN) 43(3), pp 591–604, 2009.

[DL09b] M. Dobson, M. Luskin, *An optimal order error analysis of the one-dimensional quasicontinuum approximation*, SIAM J. Numer. Anal. 47(4), pp 2455–2475, 2009.

[DLO09a] M. Dobson, M. Luskin, C. Ortner, *Sharp stability estimates for the accurate prediction of instabilities by the quasicontinuum method*, arXiv preprint 0905.2914.

[DLO09b] M. Dobson, M. Luskin, C. Ortner, *Stability, instability, and error of the force-based quasicontinuum approximation*, arXiv preprint 0903.0610.

[DS63] N. Dunford, J. T. Schwartz, *Linear operators. Volume 2: Spectral theory: self adjoint operators in Hilbert space*, Wiley, New York, 1963.

[DTMP05] L. M. Dupuy, E. B. Tadmor, R. E. Miller, R. Phillips, *Finite temperature Quasicontinuum: molecular dynamics without all the atoms*, Phys. Rev. Lett. 95, 060202, 2005.

[EM04] W. E, P. B. Ming, *Analysis of multiscale methods*, J. Comp. Math. 22(2), pp 210–219, 2004.

[EM07] W. E, P. B. Ming, *Cauchy-Born rule and stability of crystals: static problems*, Arch. Rat. Mech. Anal. 183(2), pp 241–297, 2007.

[Ell85a] R. S. Ellis, *Entropy, large deviations, and statistical mechanics*, volume 271 of Grundlehren der Mathematischen Wissenschaften. Springer-Verlag, New York, 1985.

[Ell85b] R. S. Ellis, *Large deviations and statistical mechanics.* In *Particle systems, random media and large deviations*, (Brunswick, Maine, 1984), volume 41 of Contemp. Math., pp 101–123. Amer. Math. Soc., Providence, RI, 1985.

[Ell95] R. S. Ellis, *An overview of the theory of large deviations and applications to statistical mechanics*, Scand. Actuar. J. 1, pp 97–142, 1995. Harald Cramer Symposium (Stockholm, 1993).

[dH00] F. den Hollander, *Large deviations.* Fields Institute Monographs, 14. American Mathematical Society, Providence, RI, 2000.

[KO01] J. Knap, M. Ortiz, *An analysis of the quasicontinuum method*, J. Mech. Phys. Solids 49(9), pp 1899–1923, 2001.

[KC78] H. J. Kushner, D. S. Clark, *Stochastic approximation methods for constrained and unconstrained systems*, volume 26 of Applied Mathematical Sciences. Springer-Verlag, New York, 1978.

[Leg09] F. Legoll, *Multiscale methods coupling atomistic and continuum mechanics: some examples of mathematical analysis*, in Analytical and Numerical Aspects of Partial Differential Equations, E. Emmrich and P. Wittbold eds., de Gruyter, pp 193–245 (2009).

[LSNS89] R. LeSar, R. Najafabadi, D. J. Srolovitz, *Finite-temperature defect properties from free-energy minimization*, Phys. Rev. Lett. 63, pp 624–627, 1989.

[Lin03] P. Lin, *Theoretical and numerical analysis of the quasi-continuum approximation of a material particle model*, Math. Comput. 72, pp 657–675, 2003.

[Lin07] P. Lin, *Convergence analysis of a quasi-continuum approximation for a two-dimensional material*, SIAM J. Numer. Anal. 45(1), pp 313–332, 2007.

[LT95] P.-L. Lions, G. Toscani, *A strengthened central limit theorem for smooth densities*, J. Funct. Anal. 128, pp 148–176, 1995.

[MT09] S. P. Meyn, R. L. Tweedie, *Markov chains and stochastic stability.* Springer (2009).

[MTPO98] R. Miller, E. B. Tadmor, R. Phillips, M. Ortiz, *Quasicontinuum simulation of fracture at the atomic scale*, Modelling Simul. Mater. Sci. Eng. 6, pp 607–638, 1998.

[MT02] R. Miller, E. B. Tadmor, *The quasicontinuum method: overview, applications and current directions*, Journal of Computer-Aided Materials Design 9, pp 203–239, 2002.

[Oll07] S. Olla, *Non-equilibrium macroscopic behavior of chain of interacting oscillators*, lecture notes from a course at IHP (Paris), fall 2007.

[OS08] C. Ortner, E. Süli, *Analysis of a quasicontinuum method in one dimension*, Math. Model. Numer. Anal. (M2AN) 42(1), pp 57–91, 2008.

[Pat09] C. Patz, PhD. dissertation, in preparation.

[Pre09] E. Presutti, *Scaling limits in statistical mechanics and microstructures in continuum mechanics*, Springer-Verlag, 2008.

[SW99] H. Schaefer, M. P. Wolff, *Topological vector spaces*, second edition. Graduate Texts in Mathematics, 3. Springer-Verlag, New York, 1999.

[SMT+98] V. B. Shenoy, R. Miller, E. B. Tadmor, R. Phillips, M. Ortiz, *Quasicontinuum models of interfacial structure and deformation*, Phys. Rev. Lett. 80(4), pp 742-745, 1998.

[SMT+99] V. B. Shenoy, R. Miller, E. B. Tadmor, D. Rodney, R. Phillips, M. Ortiz, *An adaptative finite element approach to atomic-scale mechanics - the quasicontinuum method*, J. Mech. Phys. Solids 47, pp 611–642, 1999.

[TSBK99] E. B. Tadmor, G. S. Smith, N. Bernstein, E. Kaxiras, *Mixed finite element and atomistic formulation for complex crystals*, Phys. Rev. B 59(1), pp 235-245, 1999.

[TPO96] E. B. Tadmor, R. Phillips, M. Ortiz *Mixed atomistic and continuum models of deformation in solids*, Langmuir 12, pp 4529–4534, 1996.

[TOP96] E. B. Tadmor, M. Ortiz, R. Phillips, *Quasicontinuum analysis of defects in solids*, Phil. Mag. A 73, pp 1529–1563, 1996.

[Var84] S. R. S. Varadhan, *Large deviations and applications*, SIAM, Philadelphia, 1984.

Bibliography

[AG05] M. Arndt and M. Griebel. Derivation of higher order gradient continuum models from atomistic models for crystalline solids. *Multiscale Model. Simul.*, 4(2), 531–562 (electronic), 2005.

[AL08] M. Arndt and M. Luskin. Error estimation and atomistic-continuum adaptivity for the quasicontinuum approximation of a Frenkel-Kontorova model. *Multiscale Model. Simul.*, 7(1), 147–170, 2008.

[AM78] R. Abraham and J. E. Marsden. *Foundations of mechanics.* Benjamin/Cummings Publishing Co. Inc. Advanced Book Program, Reading, Mass., 1978.

[ANZEA09] M. Anitescu, D. Negrut, P. Zapol, and A. El-Azab. A note on the regularity of reduced models obtained by nonlocal quasi-continuum-like approaches. *Math. Program.*, 118(2, Ser. A), 207–236, 2009.

[Bal91] R. Balian. *From microphysics to macrophysics. Vol. I.* Texts and Monographs in Physics. Springer-Verlag, Berlin, 1991. Methods and applications of statistical physics.

[Bax82] R. J. Baxter. *Exactly solved models in statistical mechanics.* Academic Press Inc. [Harcourt Brace Jovanovich Publishers], London, 1982.

[BCAH87] R. B. Bird, C. F. Curtiss, R. C. Armstrong, and O. Hassager. *Dynamics of Polymeric Liquids: Kinetic Theory.* Wiley, New York, second edition, 1987.

[BH86] N. Bleistein and R. A. Handelsman. *Asymptotic expansions of integrals.* Dover Publications Inc., New York, second edition, 1986.

[BL76] J. Bergh and J. Löfström. *Interpolation spaces. An introduction*, volume 223 of *Grundlehren der Mathematischen Wissenschaften.* Springer-Verlag, Berlin, 1976.

[BLBL02a] X. Blanc, C. Le Bris, and P.-L. Lions. From molecular models to continuum mechanics. *Arch. Ration. Mech. Anal.*, 164(4), 341–381, 2002.

[BLBL02b] X. Blanc, C. Le Bris, and P.-L. Lions. From molecular models to continuum mechanics. *Arch. Ration. Mech. Anal.*, 164(4), 341–381, 2002.

[BLBL05] X. Blanc, C. Le Bris, and F. Legoll. Analysis of a prototypical multiscale method coupling atomistic and continuum mechanics. *M2AN Math. Model. Numer. Anal.*, 39(4), 797–826, 2005.

[BLBL07a] X. Blanc, C. Le Bris, and F. Legoll. Analysis of a prototypical multiscale method coupling atomistic and continuum mechanics: the convex case. *Acta Math. Appl. Sin. Engl. Ser.*, 23(2), 209–216, 2007.

[BLBL07b] X. Blanc, C. Le Bris, and P.-L. Lions. Atomistic to continuum limits for computational materials science. *M2AN Math. Model. Numer. Anal.*, 41(2), 391–426, 2007.

[BLL75] H. J. Brascamp, E. H. Lieb, and J. L. Lebowitz. The statistical mechanics of anharmonic lattices. In *Proceedings of the 40th Session of the International Statistical Institute (Warsaw, 1975), Vol. 1. Invited papers*, volume 46, pages 393–404 (1976), 1975.

[BO78] C. M. Bender and S. A. Orszag. *Advanced mathematical methods for scientists and engineers.* McGraw-Hill Book Co., New York, 1978. International Series in Pure and Applied Mathematics.

[BP06] D. Bambusi and A. Ponno. On metastability in FPU. *Comm. Math. Phys.*, 264(2), 539–561, 2006.

[BP08] D. Bambusi and A. Ponno. Resonance, metastability and blow up in FPU. In *The Fermi-Pasta-Ulam problem*, volume 728 of *Lecture Notes in Phys.*, pages 191–205. Springer, Berlin, 2008.

[Bri53] L. Brillouin. *Wave propagation in periodic structures. Electric filters and crystal lattices.* Dover Publications Inc., New York, N. Y., second edition, 1953.

[Bri60] L. Brillouin. *Wave propagation and group velocity*, volume 8 of *Pure and Applied Physics.* Academic Press, New York, 1960.

[BRR60] R. R. Bahadur and R. Ranga Rao. On deviations of the sample mean. *Ann. Math. Statist.*, 31, 1015–1027, 1960.

[BS87] M. S. BIRMAN and M. Z. SOLOMJAK. *Spectral theory of selfadjoint operators in Hilbert space.* Mathematics and its Applications (Soviet Series). D. Reidel Publishing Co., Dordrecht, 1987.

[CC02] S. CURTAROLO and G. CEDER. Dynamics of an Inhomogeneously Coarse Grained Multiscale System. *Physical Review Letters*, 88, 2002.

[CLS07] E. CANCÈS, F. LEGOLL, and G. STOLTZ. Theoretical and numerical comparison of some sampling methods for molecular dynamics. *M2AN Math. Model. Numer. Anal.*, 41(2), 351–389, 2007.

[Con99] R. Conte, editor. *The Painlevé property.* CRM Series in Mathematical Physics. Springer-Verlag, New York, 1999. One century later.

[CW91] F. M. CHRIST and M. I. WEINSTEIN. Dispersion of small amplitude solutions of the generalized Korteweg-de Vries equation. *J. Funct. Anal.*, 100(1), 87–109, 1991.

[DEFW93] D. B. DUNCAN, J. C. EILBECK, H. FEDDERSEN, and J. A. D. WATTIS. Solitons on lattices. *Phys. D*, 68(1), 1–11, 1993.

[Der07] B. DERRIDA. Non-equilibrium steady states: fluctuations and large deviations of the density and of the current. *J. Stat. Mech. Theory Exp.*, (7), P07023, 45 pp. (electronic), 2007.

[DFP00] P. DEÁK, T. FRAUENHEIM, and M. R. PEDERSON. *Computer simulation of materials at atomic level.* Wiley, Berlin, 2000.

[dH00] F. DEN HOLLANDER. *Large deviations*, volume 14 of *Fields Institute Monographs.* American Mathematical Society, Providence, RI, 2000.

[DHM06] W. DREYER, M. HERRMANN, and A. MIELKE. Micro-macro transition in the atomic chain via Whitham's modulation equation. *Nonlinearity*, 19(2), 471–500, 2006.

[DHR06] W. DREYER, M. HERRMANN, and J. D. M. RADEMACHER. Wave trains, solitons and modulation theory in FPU chains. In *Analysis, modeling and simulation of multiscale problems*, pages 467–500. Springer, Berlin, 2006.

[DJ89] P. G. DRAZIN and R. S. JOHNSON. *Solitons: an introduction.* Cambridge Texts in Applied Mathematics. Cambridge University Press, Cambridge, 1989.

[DL08] M. DOBSON and M. LUSKIN. Analysis of a force-based quasicontinuum approximation. *M2AN Math. Model. Numer. Anal.*, 42(1), 113–139, 2008.

[DL09a] M. DOBSON and M. LUSKIN. An analysis of the effect of ghost force oscillation on quasicontinuum error. *M2AN Math. Model. Numer. Anal.*, 43(3), 591–604, 2009.

[DL09b] M. DOBSON and M. LUSKIN. An optimal order error analysis of the one-dimensional quasicontinuum approximation. *SIAM J. Numer. Anal.*, 47(4), 2455–2475, 2009.

[DLO10a] M. DOBSON, M. LUSKIN, and C. ORTNER. Accuracy of quasicontinuum approximations near instabilities. *Journal of The Mechanics and Physics of Solids*, 58, 1741–1757, 2010.

[DLO10b] M. DOBSON, M. LUSKIN, and C. ORTNER. Stability, instability, and error of the force-based quasicontinuum approximation. *Arch. Ration. Mech. Anal.*, 197(1), 179–202, 2010.

[DM72] H. DYM and H. P. MCKEAN. *Fourier series and integrals*, volume 14 of *Probability and Mathematical Statistics*. Academic Press, New York, 1972.

[DS63] N. DUNFORD and J. T. SCHWARTZ. *Linear operators. Part II: Spectral theory. Self adjoint operators in Hilbert space.* With the assistance of William G. Bade and Robert G. Bartle. Interscience Publishers John Wiley & Sons New York-London, 1963.

[DTMP05] L. M. DUPUY, E. B. TADMOR, R. E. MILLER, and R. PHILLIPS. Finite-temperature quasicontinuum: Molecular dynamics without all the atoms. *Phys. Rev. Let.*, 95(6), 060202, 2005.

[Dui73] J. J. DUISTERMAAT. *Fourier integral operators.* Courant Institute of Mathematical Sciences New York University, New York, 1973.

[Dui74] J. J. DUISTERMAAT. Oscillatory integrals, Lagrange immersions and unfolding of singularities. *Comm. Pure Appl. Math.*, 27, 207–281, 1974.

[DZ93] A. DEMBO and O. ZEITOUNI. *Large deviations techniques and applications.* Jones and Bartlett Publishers, Boston, MA, 1993.

[DZ02] A. DEMBO and O. ZEITOUNI. Large deviations and applications. In *Handbook of stochastic analysis and applications*, volume 163 of *Statist. Textbooks Monogr.*, pages 361–416. Dekker, New York, 2002.

[EF90] J. C. EILBECK and R. FLESCH. Calculation of families of solitary waves on discrete lattices. *Phys. Lett. A*, 149(4), 200–202, 1990.

[Ell85a] R. S. ELLIS. *Entropy, large deviations, and statistical mechanics*, volume 271 of *Grundlehren der Mathematischen Wissenschaften*. Springer-Verlag, New York, 1985.

[Ell85b] R. S. ELLIS. Large deviations and statistical mechanics. In *Particle systems, random media and large deviations (Brunswick, Maine, 1984)*, volume 41 of *Contemp. Math.*, pages 101–123. Amer. Math. Soc., Providence, RI, 1985.

[Ell95] R. S. ELLIS. An overview of the theory of large deviations and applications to statistical mechanics. *Scand. Actuar. J.*, (1), 97–142, 1995. Harald Cramér Symposium (Stockholm, 1993).

[EM04] W. E and P. MING. Analysis of multiscale methods. *J. Comput. Math.*, 22(2), 210–219, 2004. Special issue dedicated to the 70th birthday of Professor Zhong-Ci Shi.

[EM07] W. E and P. MING. Cauchy-Born rule and the stability of crystalline solids: static problems. *Arch. Ration. Mech. Anal.*, 183(2), 241–297, 2007.

[Fef85] C. FEFFERMAN. The thermodynamic limit for a crystal. *Comm. Math. Phys.*, 98(3), 289–311, 1985.

[Fel71] W. FELLER. *An introduction to probability theory and its applications. Vol. II.* Second edition. John Wiley & Sons Inc., New York, 1971.

[Fis64] M. E. FISHER. The free energy of a macroscopic system. *Arch. Rational Mech. Anal.*, 17, 377–410, 1964.

[FM02] G. FRIESECKE and K. MATTHIES. Atomic-scale localization of high-energy solitary waves on lattices. *Phys. D*, 171(4), 211–220, 2002.

[FM03] G. FRIESECKE and K. MATTHIES. Geometric solitary waves in a 2D mass-spring lattice. *Discrete Contin. Dyn. Syst. Ser. B*, 3(1), 105–114, 2003.

[FP99] G. FRIESECKE and R. L. PEGO. Solitary waves on FPU lattices. I. Qualitative properties, renormalization and continuum limit. *Nonlinearity*, 12(6), 1601–1627, 1999.

[FP02] G. FRIESECKE and R. L. PEGO. Solitary waves on FPU lattices. II. Linear implies nonlinear stability. *Nonlinearity*, 15(4), 1343–1359, 2002.

[FP04a] G. Friesecke and R. L. Pego. Solitary waves on Fermi-Pasta-Ulam lattices. III. Howland-type Floquet theory. *Nonlinearity*, 17(1), 207–227, 2004.

[FP04b] G. Friesecke and R. L. Pego. Solitary waves on Fermi-Pasta-Ulam lattices. IV. Proof of stability at low energy. *Nonlinearity*, 17(1), 229–251, 2004.

[FPU55] E. Fermi, J. Pasta, and S. Ulam. Studies of nonlinear problems. *Los Alamos Scientific Laboratory of the University of California*, Report LA-1940, 1955.

[Fri03] G. Friesecke. Dynamics of the infinite harmonic chain: conversion of coherent initial data into synchronized binary oscillations. *Preprint*, 2003.

[FT02] G. Friesecke and F. Theil. Validity and failure of the Cauchy-Born hypothesis in a two-dimensional mass-spring lattice. *J. Nonlinear Sci.*, 12(5), 445–478, 2002.

[FV99] A.-M. Filip and S. Venakides. Existence and modulation of traveling waves in particle chains. *Comm. Pure Appl. Math.*, 52(6), 693–735, 1999.

[FW94] G. Friesecke and J. A. D. Wattis. Existence theorem for solitary waves on lattices. *Comm. Math. Phys.*, 161(2), 391–418, 1994.

[GHM06] J. Giannoulis, M. Herrmann, and A. Mielke. Continuum descriptions for the dynamics in discrete lattices: derivation and justification. In *Analysis, modeling and simulation of multiscale problems*, pages 435–466. Springer, Berlin, 2006.

[GHM08] J. Giannoulis, M. Herrmann, and A. Mielke. The nonlinear Schrödinger equation as a macroscopic limit for an oscillator chain with cubic nonlinearities. *J. Math. Phys.*, ???, 2008.

[GM04] J. Giannoulis and A. Mielke. The nonlinear Schrödinger equation as a macroscopic limit for an oscillator chain with cubic nonlinearities. *Nonlinearity*, 17(2), 551–565, 2004.

[GM06] J. Giannoulis and A. Mielke. Dispersive evolution of pulses in oscillator chains with general interaction potentials. *Discrete Contin. Dyn. Syst. Ser. B*, 6(3), 493–523 (electronic), 2006.

[GWF81] A. D. Gorman, R. Wells, and G. N. Fleming. Wave propagation and thom's theorem. *J. Phys. A: Math. Gen.*, 14(7), 1519–1531, 1981.

[HGM34] K. F. HERZFELD and M. GOEPPERT-MAYER. On the state of aggregation. *Journ. chem. phys.*, 2, 38–45, 1934.

[HLTT08] L. HARRIS, J. LUKKARINEN, S. TEUFEL, and F. THEIL. Energy transport by acoustic modes of harmonic lattices. *SIAM J. Math. Anal.*, 40(4), 1392–1418, Jan. 2008.

[HLW02] E. HAIRER, C. LUBICH, and G. WANNER. *Geometric numerical integration*, volume 31 of *Springer Series in Computational Mathematics*. Springer-Verlag, Berlin, 2002. Structure-preserving algorithms for ordinary differential equations.

[HMB89] D. HOCHSTRASSER, F. G. MERTENS, and H. BÜTTNER. An iterative method for the calculation of narrow solitary excitations on atomic chains. *Phys. D*, 35(1-2), 259–266, 1989.

[Hol95] M. H. HOLMES. *Introduction to perturbation methods*, volume 20 of *Texts in Applied Mathematics*. Springer-Verlag, New York, 1995.

[Hon02] J. HONERKAMP. *Statistical physics*. Advanced Texts in Physics. Springer-Verlag, Berlin, second edition, 2002. An advanced approach with applications, Web-enhanced with problems and solutions.

[Hör90] L. HÖRMANDER. *The analysis of linear partial differential operators. I*, volume 256 of *Grundlehren der Mathematischen Wissenschaften*. Springer-Verlag, Berlin, second edition, 1990.

[Ign07] L. I. IGNAT. Fully discrete schemes for the Schrödinger equation. Dispersive properties. *Math. Models Methods Appl. Sci.*, 17(4), 567–591, 2007.

[IJ05] G. IOOSS and G. JAMES. Localized waves in nonlinear oscillator chains. *Chaos*, 15(1), 015113–+, March 2005.

[IKSY91] K. IWASAKI, H. KIMURA, S. SHIMOMURA, and M. YOSHIDA. *From Gauss to Painlevé*. Aspects of Mathematics, E16. Friedr. Vieweg & Sohn, Braunschweig, 1991. A modern theory of special functions.

[Ioo00] G. IOOSS. Travelling waves in the Fermi-Pasta-Ulam lattice. *Nonlinearity*, 13(3), 849–866, 2000.

[IZ05a] L. I. IGNAT and E. ZUAZUA. Dispersive properties of a viscous numerical scheme for the Schrödinger equation. *C. R. Math. Acad. Sci. Paris*, 340(7), 529–534, 2005.

[IZ05b] L. I. IGNAT and E. ZUAZUA. A two-grid approximation scheme for nonlinear Schrödinger equations: dispersive properties and convergence. *C. R. Math. Acad. Sci. Paris*, 341(6), 381–386, 2005.

[JJ56] H. JEFFREYS and B. S. JEFFREYS. *Methods of mathematical physics*. Cambridge University Press, 1956. 3d ed.

[KC78] H. J. KUSHNER and D. S. CLARK. *Stochastic approximation methods for constrained and unconstrained systems*, volume 26 of *Applied Mathematical Sciences*. Springer-Verlag, New York, 1978.

[KKR04] Y. A. KOSEVICH, R. KHOMERIKI, and S. RUFFO. Supersonic discrete kink-solitons and sinusoidal patterns with “magic” wave number in anharmonic lattices. *EPL (Europhysics Letters)*, 66(1), 21–27, 2004.

[KO01] J. KNAP and M. ORTIZ. An analysis of the quasicontinuum method. *Journal of The Mechanics and Physics of Solids*, 49, 1899–1923, 2001.

[Leg09] F. LEGOLL. Multiscale methods coupling atomistic and continuum mechanics: some examples of mathematical analysis. In *Analytical and numerical aspects of partial differential equations*, pages 193–245. Walter de Gruyter, Berlin, 2009.

[Len73] A. Lenard, editor. *Statistical mechanics and mathematical problems*. Springer-Verlag, Berlin, 1973. Battelle Rencontres, Seattle, Wash., 1971, Lecture Notes in Physics, Vol. 20.

[Lie76] E. H. LIEB. The stability of matter. *Rev. Modern Phys.*, 48(4), 553–569, 1976.

[Lie90] E. H. LIEB. The stability of matter: from atoms to stars. *Bull. Amer. Math. Soc. (N.S.)*, 22(1), 1–49, 1990.

[Lie91] E. H. LIEB. *The stability of matter: From atoms to stars*. Springer-Verlag, Berlin, 1991. Selecta of Elliott H. Lieb, Edited by W. Thirring, with a preface by F. Dyson.

[Lie04] E. H. LIEB. *Statistical mechanics*. Springer-Verlag, Berlin, 2004. Selecta of Elliott H. Lieb, Edited, with a preface and commentaries, by B. Nachtergaele, J. P. Solovej and J. Yngvason.

[Lin03] P. LIN. Theoretical and numerical analysis for the quasi-continuum approximation of a material particle model. *Math. Comp.*, 72(242), 657–675 (electronic), 2003.

[Lin07] P. LIN. Convergence analysis of a quasi-continuum approximation for a two-dimensional material without defects. *SIAM J. Numer. Anal.*, 45(1), 313–332 (electronic), 2007.

[LL69] E. H. LIEB and J. L. LEBOWITZ. Existence of thermodynamics for real matter with couloumb forces. *Phys. Rev. Lett.*, 22(13), 631–634, 1969.

[LL72] E. H. LIEB and J. L. LEBOWITZ. The constitution of matter: Existence of thermodynamics for systems composed of electrons and nuclei. *Advances in Math.*, 9, 316–398, 1972.

[LL75] O. E. LANFORD, III and J. L. LEBOWITZ. Time evolution and ergodic properties of harmonic systems. In *Dynamical systems, theory and applications (Rencontres, Battelle Res. Inst., Seattle, Wash., 1974)*, pages 144–177. Lecture Notes in Phys., Vol. 38. Springer, Berlin, 1975.

[LM66] E. H. LIEB and D. MATTIS. *Mathematical physics in one dimension.* Academic Press, New York, 1966.

[LNS89] R. LESAR, R. NAJAFABADI, and D. J. SROLOVITZ. Finite-temperature defect properties from free-energy minimization. *Physical Review Letters*, 63, 624–627, 1989.

[LT95] P.-L. LIONS and G. TOSCANI. A strengthened central limit theorem for smooth densities. *J. Funct. Anal.*, 129(1), 148–167, 1995.

[Mac04] F. MACIÀ. Wigner measures in the discrete setting: high-frequency analysis of sampling and reconstruction operators. *SIAM J. Math. Anal.*, 36(2), 347–383 (electronic), 2004.

[McM02] E. MCMILLAN. Multiscale correction to solitary wave solutions on FPU lattices. *Nonlinearity*, 15(5), 1685–1697, 2002.

[McM05] E. MCMILLAN. On the impact of FPU lattice discreteness upon solitary wave interactions. *Phys. D*, 200(1-2), 25–46, 2005.

[Mie04] A. MIELKE. Macroscopic behavior of microscopic oscillations in harmonic lattices. *Preprint of DFG Priority Program Analysis, Modeling and Simulation of Multiscale Problems*, 118, 2004.

[Mie06a] A. Mielke, editor. *Analysis, modeling and simulation of multiscale problems.* Springer-Verlag, Berlin, 2006.

[Mie06b] A. MIELKE. Macroscopic behavior of microscopic oscillations in harmonic lattices via Wigner-Husimi transforms. *Arch. Ration. Mech. Anal.*, 181(3), 401–448, 2006.

[Mil06] P. D. MILLER. *Applied asymptotic analysis*, volume 75 of *Graduate Studies in Mathematics.* American Mathematical Society, Providence, RI, 2006.

[Mon56] E. W. Montroll. Theory of the vibration of simple cubic lattices with nearest neighbor interactions. In *Proceedings of the Third Berkeley Symposium on Mathematical Statistics and Probability, 1954–1955, vol. III*, pages 209–246, Berkeley and Los Angeles, 1956. University of California Press.

[MT02] R. Miller and E. B. Tadmor. The quasicontinuum method: Overview, applications and current directions. *J. Comput. Aided Mater. Des.*, 9(3), 203–239, 2002.

[MT09] S. Meyn and R. L. Tweedie. *Markov chains and stochastic stability.* Cambridge University Press, Cambridge, second edition, 2009. With a prologue by Peter W. Glynn.

[MTPO98] R. Miller, E. B. Tadmor, R. Phillips, and M. Ortiz. Quasicontinuum simulation of fracture at the atomic scale. *Modelling and Simulation in Materials Science and Engineering*, 6, 607–638, 1998.

[Mur84] J. D. Murray. *Asymptotic analysis*, volume 48 of *Applied Mathematical Sciences.* Springer-Verlag, New York, second edition, 1984.

[Nag40a] T. Nagamiya. Statistical mechanics of one-dimensional substances. I. *Proc. Phys.-Math. Soc. Japan (3)*, 22, 705–720, 1940.

[Nag40b] T. Nagamiya. Statistical mechanics of one-dimensional substances. II. *Proc. Phys.-Math. Soc. Japan (3)*, 22, 1034–1047, 1940.

[Olv74] F. W. J. Olver. *Asymptotics and special functions.* Academic Press, New York-London, 1974. Computer Science and Applied Mathematics.

[OS08] C. Ortner and E. Süli. Analysis of a quasicontinuum method in one dimension. *M2AN Math. Model. Numer. Anal.*, 42(1), 57–91, 2008.

[Paz83] A. Pazy. *Semigroups of linear operators and applications to partial differential equations*, volume 44 of *Applied Mathematical Sciences.* Springer-Verlag, New York, 1983.

[Pre09] E. Presutti. *Scaling limits in statistical mechanics and microstructures in continuum mechanics.* Theoretical and Mathematical Physics. Springer, Berlin, 2009.

[PS72] O. Penrose and E. R. Smith. Thermodynamic limit for classical systems with Coulomb interactions in a constant external field. *Comm. Math. Phys.*, 26, 53–77, 1972.

[Rac92] R. RACKE. *Lectures on nonlinear evolution equations*, volume E19 of *Aspects of Mathematics*. Friedr. Vieweg & Sohn, Braunschweig, 1992.

[Ree76] M. REED. *Abstract non-linear wave equations*. Lecture Notes in Mathematics, Vol. 507. Springer-Verlag, Berlin, 1976.

[RH93] P. ROSENAU and J. M. HYMAN. Compactons: Solitons with finite wavelength. *Phys. Rev. Lett.*, 70(5), 564–567, Feb 1993.

[Ros86] P. ROSENAU. Dynamics of nonlinear mass-spring chains near the continuum limit. *Phys. Lett. A*, 118(5), 222–227, 1986.

[Ros87] P. ROSENAU. Dynamics of dense lattices. *Phys. Rev. B (3)*, 36(11), 5868–5876, 1987.

[Ros94] P. ROSENAU. Nonlinear dispersion and compact structures. *Phys. Rev. Lett.*, 73(13), 1737–1741, Sep 1994.

[Ros03] P. ROSENAU. Hamiltonian dynamics of dense chains and lattices: or how to correct the continuum. *Phys. Lett. A*, 311(1), 39–52, 2003.

[Rue63] D. RUELLE. Classical statistical mechanics of a system of particles. *Helv. Phys. Acta*, 36, 183–197, 1963.

[Rue69] D. RUELLE. *Statistical mechanics: Rigorous results*. W. A. Benjamin, Inc., New York-Amsterdam, 1969.

[Sch02] F. SCHWABL. *Statistical mechanics*. Advanced Texts in Physics. Springer-Verlag, Berlin, 2002. Translated from the 2000 German original by William Brewer.

[Sco03] A. SCOTT. *Nonlinear science*, volume 8 of *Oxford Texts in Applied and Engineering Mathematics*. Oxford University Press, Oxford, second edition, 2003. Emergence and dynamics of coherent structures.

[Seg68] I. SEGAL. Dispersion for non-linear relativistic equations. II. *Ann. Sci. École Norm. Sup. (4)*, 1, 459–497, 1968.

[Shi84] A. N. SHIRYAYEV. *Probability*, volume 95 of *Graduate Texts in Mathematics*. Springer-Verlag, New York, 1984. Translated from the Russian by R. P. Boas.

[SK05] A. STEFANOV and P. G. KEVREKIDIS. Asymptotic behaviour of small solutions for the discrete nonlinear Schrödinger and Klein-Gordon equations. *Nonlinearity*, 18(4), 1841–1857, 2005.

[SMT+98] V. B. SHENOY, R. MILLER, E. B. TADMOR, R. PHILLIPS, and M. ORTIZ. Quasicontinuum Models of Interfacial Structure and Deformation. *Physical Review Letters*, 80, 742–745, 1998.

[SMT$^+$99] V. B. SHENOY, R. MILLER, E. B. TADMOR, D. RODNEY, R. PHILLIPS, and M. ORTIZ. An adaptive finite element approach to atomic-scale mechanics—the quasicontinuum method. *J. Mech. Phys. Solids*, 47(3), 611–642, 1999.

[Spo91] H. SPOHN. *Large Scale Dynamics of Interacting Particles.* Texts and Monographs in Physics. Springer-Verlag, Heidelberg, 1991.

[Ste93] E. M. STEIN. *Harmonic analysis: real-variable methods, orthogonality, and oscillatory integrals*, volume 43 of *Princeton Mathematical Series.* Princeton University Press, Princeton, NJ, 1993.

[Str74] W. A. STRAUSS. Dispersion of low-energy waves for two conservative equations. *Arch. Rational Mech. Anal.*, 55, 86–92, 1974.

[Str78] W. A. STRAUSS. Nonlinear invariant wave equations. In *Invariant wave equations (Proc. "Ettore Majorana" Internat. School of Math. Phys., Erice, 1977)*, volume 73 of *Lecture Notes in Phys.*, pages 197–249. Springer, Berlin, 1978.

[SW71] E. M. STEIN and G. WEISS. *Introduction to Fourier analysis on Euclidean spaces*, volume 32 of *Princeton Mathematical Series.* Princeton University Press, Princeton, N.J., 1971.

[SW99] H. H. SCHAEFER and M. P. WOLFF. *Topological vector spaces*, volume 3 of *Graduate Texts in Mathematics.* Springer-Verlag, New York, second edition, 1999.

[SW00] G. SCHNEIDER and C. E. WAYNE. Counter-propagating waves on fluid surfaces and the continuum limit of the Fermi-Pasta-Ulam model. In B. Fiedler, K. Gröger, and J. Sprekels, editors, *International Conference on Differential Equations*, volume 1, pages 390–404. World Scientific, 2000.

[Tak42] H. TAKAHASI. Eine einfache Methode zur Behandlung der statistischen Mechanik eindimensionaler Substanzen. *Proc. Phys.-Math. Soc. Japan (3)*, 24, 60–62, 1942.

[Tao06] T. TAO. *Nonlinear dispersive equations*, volume 106 of *CBMS Regional Conference Series in Mathematics.* Published for the Conference Board of the Mathematical Sciences, Washington, DC, 2006. Local and global analysis.

[Tay11] M. E. TAYLOR. *Partial differential equations I. Basic theory*, volume 115 of *Applied Mathematical Sciences.* Springer, New York, second edition, 2011.

[Tem97] R. TEMAM. *Infinite-dimensional dynamical systems in mechanics and physics*, volume 68 of *Applied Mathematical Sciences*. Springer-Verlag, New York, second edition, 1997.

[Tho88] C. J. THOMPSON. *Classical equillibrium statistical mechanics*. Oxford University Press, Oxford, 1988.

[Ton36] L. TONKS. The complete equation of state of one, two and three-dimensional gases of hard elastic spheres. *Physical Rev.*, 50, 955–963, 1936.

[TOP96] E. B. TADMOR, M. ORTIZ, and R. PHILLIPS. Quasicontinuum analysis of defects in solids. *Philosophical Magazine A-physics of Condensed Matter Structure Defects and Mechanical Properties*, 73, 1529–1563, 1996.

[TPO96] E. B. TADMOR, R. PHILLIPS, and M. ORTIZ. Mixed Atomistic and Continuum Models of Deformation in Solids. *Langmuir*, 12, 4529–4534, 1996.

[TSBK99] E. B. TADMOR, G. S. SMITH, N. BERNSTEIN, and E. KAXIRAS. Mixed finite element and atomistic formulation for complex crystals. *Physical Review B*, 59, 235–245, 1999.

[Var84] S. R. S. VARADHAN. *Large deviations and applications*, volume 46 of *CBMS-NSF Regional Conference Series in Applied Mathematics*. Society for Industrial and Applied Mathematics (SIAM), Philadelphia, PA, 1984.

[vH49] L. VAN HOVE. Quelques propriétés générales de l'intégral de configuration d'un système de particules avec interaction. *Physica*, 15, 951–961, 1949.

[vH50] L. VAN HOVE. Sur l'intégrale de configuration pour les systèmes de particules à une dimension. *Physica*, 16, 137–143, 1950.

[Wat95] G. N. WATSON. *A treatise on the theory of Bessel functions*. Cambridge Mathematical Library. Cambridge University Press, Cambridge, 1995. Reprint of the second edition, 1944.

[Whi74] G. B. WHITHAM. *Linear and nonlinear waves*. Wiley-Interscience [John Wiley & Sons], New York, 1974. Pure and Applied Mathematics.

[Won89] R. WONG. *Asymptotic approximations of integrals*. Computer Science and Scientific Computing. Academic Press Inc., Boston, MA, 1989.

[WW02] C. E. WAYNE and J. D. WRIGHT. Higher order modulation equations for a Boussinesq equation. *SIAM J. Appl. Dyn. Syst.*, 1(2), 271–302 (electronic), 2002.

[ZK65] N. J. ZABUSKY and M. D. KRUSKAL. Interaction of "solitons" in a collisionless plasma and the recurrence of initial states. *Phys. Rev. Lett.*, 15(6), 240–243, Aug 1965.